GW01607352

THE PORSCHE DRIVING BOOK

THE PORSCHE DRIVING BOOK

Martin Beck-Burridge
John Lyon

PORSCHE CARS GREAT BRITAIN LIMITED

First Published in 1988

British Library Cataloguing in Publication Data

Beck-Burridge, Martin, *1941–*
The Porsche driving book.
1. Cars. Driving – Manuals
I. Title II. Lyon, John, *1938–*
629.28′32

ISBN 0-9513737-0-6

Published by
Porsche Cars Great Britain Limited
Bath Road, Calcot, Reading, Berkshire RG3 7SE

Design by Behram Kapadia

Photographs by Richard Newton, Behram Kapadia, Bill Bates and Porsche AG
Cover photograph by Mike Valente

Printed in Great Britain by
BAS Printers, Over Wallop, Stockbridge, Hampshire SO20 8JD

Contents

Acknowledgements

The authors are grateful for the support of Porsche Cars Great Britain Limited in commissioning this book. Many individuals in the Porsche "family" gave freely of their time and expertise. In particular Rolf Hannes, Klaus Parr, Corinna Phillips and Melanie Hill all responded in true Porsche fashion to requests for information, photographs and loan of cars. Michael Cotton and Chief Inspector Fred Ashmore's prompt comments on the manuscript were invaluable. Lastly, but most importantly, our thanks to Jeremy Snook for his meticulous enthusiasm and commitment, without which the project would not have been possible.

Naturally all errors or omissions are the authors' responsibility.

FOREWORD

Ever since the first car to bear the Porsche name was built back in 1948, we at Porsche have sought to produce cars for discerning individuals who desired above all a car with superior handling that was swift, safe and intensely satisfying to drive.

In our constant pursuit of "Driving in its purest form", we have never lost sight of one of the essential ingredients - safety, both passive and active.

This constant attention to safety is supplemented worldwide by practical training on our customer driving days. These days are designed not only to provide customers with a finer appreciation of the handling qualities of their Porsche cars but to educate them on safe performance driving techniques in a closed circuit environment.

However, by their very nature, these days can necessarily only satisfy a proportion of Porsche customers and I therefore welcome the initiative of our wholly owned subsidiary, Porsche Cars Great Britain Limited, in producing this book on advanced driving technique aimed specifically at the Porsche driver.

Professor Dr. Ferdinand Porsche.

INTRODUCTION

Driving Porsches – Driving in its Purest Form

Porsche high performance road cars are the epitome of performance engineering and have always been designed and built to give you, the driver, "Driving in its purest form". The company's first race success came after only a month's production when a private owner, Herbert Kaes, entered a Porsche 356 in a street race in Innsbruck on 11 July, 1948 and easily won his class. Since that modest beginning, Porsche cars have won most of motor racing's glittering prizes including 12 outright wins at Le Mans (a record for a single car marque), the Monte Carlo Rally, Can Am and World Sports Car Championships. Their consistent record of success is remarkable for any marque, but particularly so for a company which is one of the world's smallest motor car manufacturers, but the largest manufacturer of high performance road cars.

The company's founder, Professor Dr. Ferdinand Porsche was born in 1875. Educated in Vienna he rapidly established an outstanding reputation as a brilliant and innovative engineer. At the age of 24 he designed the Lohner-Porsche, the first four-wheel drive car, shown at the Paris Exhibition of 1900. Throughout the next fifty years, his influence upon the developing automobile industry was outstanding. As Managing Director of Austro-Daimler, he designed several motor cars including a car remarkable for its size and light weight, the 'Sascha', which won its class in the 1922 Targa Florio. In 1923 he became Technical Director of Daimler Motoren responsible for the mighty SS and SSK Mercedes Benz sports cars, and remained there until he joined the firm of Steyr in 1928. But large corporations and creative genius often make uncomfortable partners and so, after more than a quarter of a century in the German automobile industry Ferdinand Porsche established his own design company Dr. Ing. h. c. F. Porsche Gmbh on 25 April, 1931.

The German automobile industry was soon beating a path to the door of the new design consultancy. In 1932 it was commissioned to design a Grand Prix racing car for a consortium of four companies, Audi, DKW, Horch and Wanderer. The Auto-Union P-wagen was a complete break from traditional racing car design of the period and was almost thirty years ahead of its time. The driver sat very low,

'Ferry' left, and his father with the first Porsche 356, at Gmünd on July 8th, 1948.

well in front of the rear-mounted engine. This magnificent racing car included many other innovations such as swing axle rear and torsion bar front suspension, a unique valve operating mechanism, and a five-speed gearbox. Furthermore it was the first racing car to have a limited-slip differential. The Auto-Union was undoubtedly a difficult car to handle with an excessive amount of oversteer and enough power to spin its rear wheels on a dry road from 86 mph! But its successes against the finest Grand Prix cars of the day proved the brilliance of the design genius of the tin smith's son from Austria. On October 26, 1937, Bernd Rosemeyer drove the Auto-Union P-car at an average of 251.2 mph over 5 kilometres and then set two world records for the standing kilometre and the mile; records which stood until 1958. Between 1934 and 1939 this extraordinary racing car won almost 40% of its races, a total of 24 Grand Prix.

At the Berlin Automobile Show of 1934 the government announced that the German automobile industry was expected to build a practical and economical small four-seater saloon. The requirements were strict but simple; "a car priced low enough to be bought by anyone with enough funds to purchase a motorcycle. Few parts to go wrong. Low repair costs . A real people's car – a sort of Volkswagen you might say." The Porsche company already had appropriate design ideas based on projects undertaken for Zündapp and N.S.U. Furthermore it benefited from the inter-company rivalry which then ensued between all the major German manufacturers

The extremely successful 16-cylinder Auto-Union P-Wagen, Grand Prix racing car, designed by Ferdinand Porsche. In this car Bernd Rosemeyer became the first man to travel at over 250 mph on a normal road, and established standing start kilometre and mile records that stood for over 20 years.

(Opposite above) Professor Dr. Ferdinand Porsche with his Volkswagen design.

(Opposite below) The Type 360 Cisitalia Grand Prix racing car designed by 'Ferry' Porsche in 1947. This car had a 12-cylinder 1493 cc engine, a five-speed synchromesh gearbox and four-wheel drive.

who refused to cooperate with one another. The Porsche concept was, as usual, advanced and included independent suspension, no chassis but a rigid steel floor, a flat-four air-cooled 'boxer' engine with high power output at low revolutions for longer life, and a rear engine layout. The price limit of 1000 Reichmarks was rather more difficult to achieve. By 1936, more than thirty prototypes had been built and successfully tested. Even though the outbreak of hostilities in 1939 prevented the car going into production until after the war, it was to become the most produced car of all time and is still made under licence in Brazil. As with almost all of Porsche's designs this car has a devoted and enthusiastic following.

The Second World War stopped all further design developments on civilian motorcars. The company resumed work on civilian projects in 1946 under the determined and inspired leadership of Professor Dr. Porsche's son "Ferry" who was to rebuild the company created by his father, and in which he had worked all his life, into the world's finest manufacturer of high performance road cars. In that year the Managing Director of the Cisitalia works in Turin commissioned Porsche to design a Grand Prix car and Ferry and his Chief Engineer Rabe, despite the absence of Professor Dr. Ferdinand (then interned in France), completed the design of a revolutionary Grand Prix car with a 1500 cc engine in August, 1947. The car had a number of features familiar to the modern motorist; four-wheel drive and a five-speed baulk ring synchronised gearbox. The design layout

included, of course, the engine, clutch and gearbox in front of the rear axle. Tragically, the car was never to race in Europe, as Cisitalia went bankrupt.

Born in 1909, Ferry Porsche had already proved to be a remarkable engineer, but rebuilding the company required exceptional commercial flair and judgement. He continued to build on the engineering legend created by his father, but it was his drive and imagination that placed the Porsche badge on the first of a range of superb high performance road cars, the Porsche 356. So numbered because that was its project number, this car's design origins were rooted in the Auto Union P-Wagen and the Volkswagen Beetle. The car had a steel platform chassis, an aluminium body and a rear mounted four-cylinder 1,131 cc VW engine, 'turned around' so that it was in the middle. The first real Porsche , the 356 was designed and developed by Ferry Porsche, Karl Rabe and Erwin Kommenda from design studies based upon the Volkswagen Beetle but shelved in 1939. A very attractive design, the car's performance rapidly established the company's reputation as a constructor of superb high performance road cars that its enthusiastic owners could race successfully. Professor Dr. Ferdinand Porsche died on 30 January, 1951, and thus did not see see his company's first international race success, a class win in the 1951 Le Mans 24 Hours and more than 1,000 Porsche 356 cars sold to a worldwide market of enthusiasts.

The Porsche 356 originally had the crash gearbox from the VW Beetle but the baulk-ring synchromesh gearbox designed for the Cisitalia project in 1946 was far superior to anything available then, or since. In June, 1952, Porsche introduced the first production car to carry this mechanism which has since been adopted under licence as the standard form of synchromesh by many of the world's automobile manufacturers. By the end of 1965, over 76,000 Porsche 356s had been sold and made Porsche a worldwide trademark. The 356 retains a loyal following and many are still lovingly maintained and used. Under Porsche's leadership, the company continued to increase its reputation as the producer of cars for drivers "who desired above all a car with superior handling that was swift, safe and intensely satisfying to drive". The company went from strength to strength and in 1956 work started on the replacement for the 356. As the company's Chief Stylist, Ferry Porsche's son Ferdinand, nicknamed 'Butzi', was responsible for the creation of the successor to the 356 which was to become the 'classic' Porsche – the 911. Launched at the Frankfurt Motor Show in 1963, the final body shape was almost exactly as 'Butzi' had drawn it although during development several attempts were made to stretch the design to a four-seater. These views were not accepted and the company's design philosophy has since been to produce only high performance road cars of a two plus two design. The engineering ancestry of his grandfather's Auto-Union Grand Prix car and the Volkswagen Beetle are obvious in this classic

Car and designer. 'Butzi' Porsche and the 1963 Porsche Type 901(T8), with the 2-litre flat six-cylinder engine.

amongst the world's great supercars. In a quarter of a century of continuous development the Porsche 911 has remained a true high performance road car and many regard it as the finest sports car the world has ever seen. The 911 Turbo is the only supercar that seems as comfortable when driven in the capitals of the world as it is either covering continents at high, safe speeds or competing successfully on the world's race tracks.

The pursuit of excellence is neither quick nor cheap, success is an ever moving target and only persistent attention to detail wins; either on the road or race track. Porsche has never wanted, nor ever been big enough, to simply 'follow' fashion. Its cars have all been based on a brilliant initial concept which has been continuously developed upon sound engineering principles.

During the late 1960s the Porsche design team started to develop its ideas for the next generation of cars. A new motorcar design now requires massive resources to bring it to the production stage and Porsche has never had the resources of the mass producers. All Porsche models have to have a 'life' of at least a dozen years; the Porsche 911 is a classic in that sense too. The company, in spite of the cold wind of the oil crisis, was determined to build a new series of cars for the late twentieth century. The danger was that marketing and sales thinking would dominate the situation rather than driver enthusiasts who were also engineers; they were the people who knew and identified with Porsche's market.

But the controlling families of Porsche had further problems to face, that of too much family talent. Professor Dr. Ferdinand Porsche's two children, Ferdinand (Ferry) and Louise, both had talented children. Louise's son, Ferdinand Piëch was Porsche's top engineer and Butzi (Ferdinand's son) was chief stylist. There were simply too many talented young 'turks', all convinced that they should be running the company. Dr. Ferdinand Porsche's solution was to remove all the children from the company's executive management and for the family to become stockholding overseers. In 1972, both of the owning families stepped down from executive control and a new executive management board was appointed under the leadership of Dr Ernst Führmann. Butzi Porsche set up his own very successful design studio and Ferdinand Piëch took the top engineering job at Audi and now heads that company.

The transition from being a wholly owned family firm, to independent board control, and the realisation that the company needed to re-examine and perhaps extend its range of cars, happened practically simultaneously. Determined that it would continue to produce the world's most desired sports coupés, the management knew that the new Research and Development Centre at Weissach would ensure that the company remained at the forefront of automotive engineering. The next major step was the development of a new generation of Porsche high performance road cars which would ensure that the company retained its pre-eminent position in the world's markets. The first of the new cars came to production via Volkswagen; and somewhat by chance.

In 1970, Porsche had been entrusted with the development of project EA425 by Volkswagenwerk. Designated as a replacement for the VW – Porsche 914, it was cancelled by VW's new Managing Director. The car was by then almost fully developed and Porsche decided to produce the car in its own production programme at the Audi – N.S.U. factory at Neckarsulm. The car was of the same design layout as the 928 then under development, but there the similarity ended. All of these new Porsches were a radical departure from the 'classic' Porsche layout, with a water-cooled engine and clutch in the front , and a combined gearbox and drive unit in the rear. This Transaxle layout gives an almost perfect 49% front / 51% rear weight distribution that endows all of these models with particularly predictable handling characteristics. The first 'new generation' Porsche to be launched was the 924 in 1975, and within a year it had become the world's best selling sports car.

Following the success of the new car, the company launched two further models in the range, including its new 'flagship', the Porsche 928. Launched in 1977, this car had a 4.5-litre V8 engine and several major innovations and, in 1978, became the first and only sports car to win 'The Car of the Year' award. Then in 1981, the Le Mans 24-Hour race was won outright by a Porsche 936 with a Porsche

'The Car of the Year' in 1978. The Porsche 928 set new standards of comfort and performance.

924 GTP in seventh place. This latter car was the racing prototype of the Porsche 944 launched in 1981, with a new 2479 cc water-cooled four-cylinder engine.

The ultimate Porsche, at the moment is the 959. A perfect example of Porsche's philosophy of continuous development, this stunning example of ultra-high performance automotive engineering is the latest development of the 911, and has all (and more), of the superb handling qualities of that classic of the marque. But there the similarity ends as this car owes few components to previous models. A unique electronically-optimised four-wheel drive system, body panels made from Kevlar and other new materials, self-levelling suspension, literally unrivalled performance from a 450bhp 2.85-litre engine, plus typical Porsche manners make the 959 the ultimate supercar in the world today.

The original company was a design, research and development consultancy and Porsche has always recognised the need for continuous research and development effort. During the last two decades the company's Research and Development Centre at Weissach has become an important contributor to the company's financial profitability as well as its prominence in all fields of engineering. The projects which have emerged from this remarkably fertile engineering

Powered by the TAG Tubo engine made by Porsche, the McLaren Grand Prix car won three World Drivers' Championships and two Constructors' Championships.

(Opposite above) The PFM 3200 aero-engine. Developed from the six-cylinder, air-cooled 911 engine, this is the quietest and most powerful aero-engine in its class.

(Opposite below) Powered by the Porsche 3200 aero-engine this Mooney made over 300 take-offs and landings on a problem free round the world proving flight.

power-house cover practically the whole engineering industry. Of the company's approximately 8,500 employees some 26% work at Weissach, but their contribution in profits and engineering terms is, to say the least, impressive in its range and excellence.

The 1.5-litre turbocharged TAG Turbo Formula One engine made by Porsche dominated the Grand Prix racing scene in 1984, 1985 and 1986. Such excellence in its own field is expected but the company has also produced the A310 Airbus cockpit, the PFM 3200 air-cooled six-cylinder aero-engine and a remarkably versatile tank retriever that has proved useful in many civilian catastrophes. All this and many other outside contracts serve to emphasise the company's continued technological excellence which ensures that its products are at the forefront of automotive design. Concern for speed, safety and handling is not all and the Longlife car designed in 1973 shows that the company has an impressive stake in the application of new materials to chassis and engine design and the development of economical solutions to urban transportation. Never a company to waste ideas and always more concerned with development than fashion, the Longlife vehicle proved to Porsche the value of galvanising body panels. Since 1976 all Porsche panels have been so treated and now carry a 10-year routine maintenance free Longlife anti-corrosion body warranty.

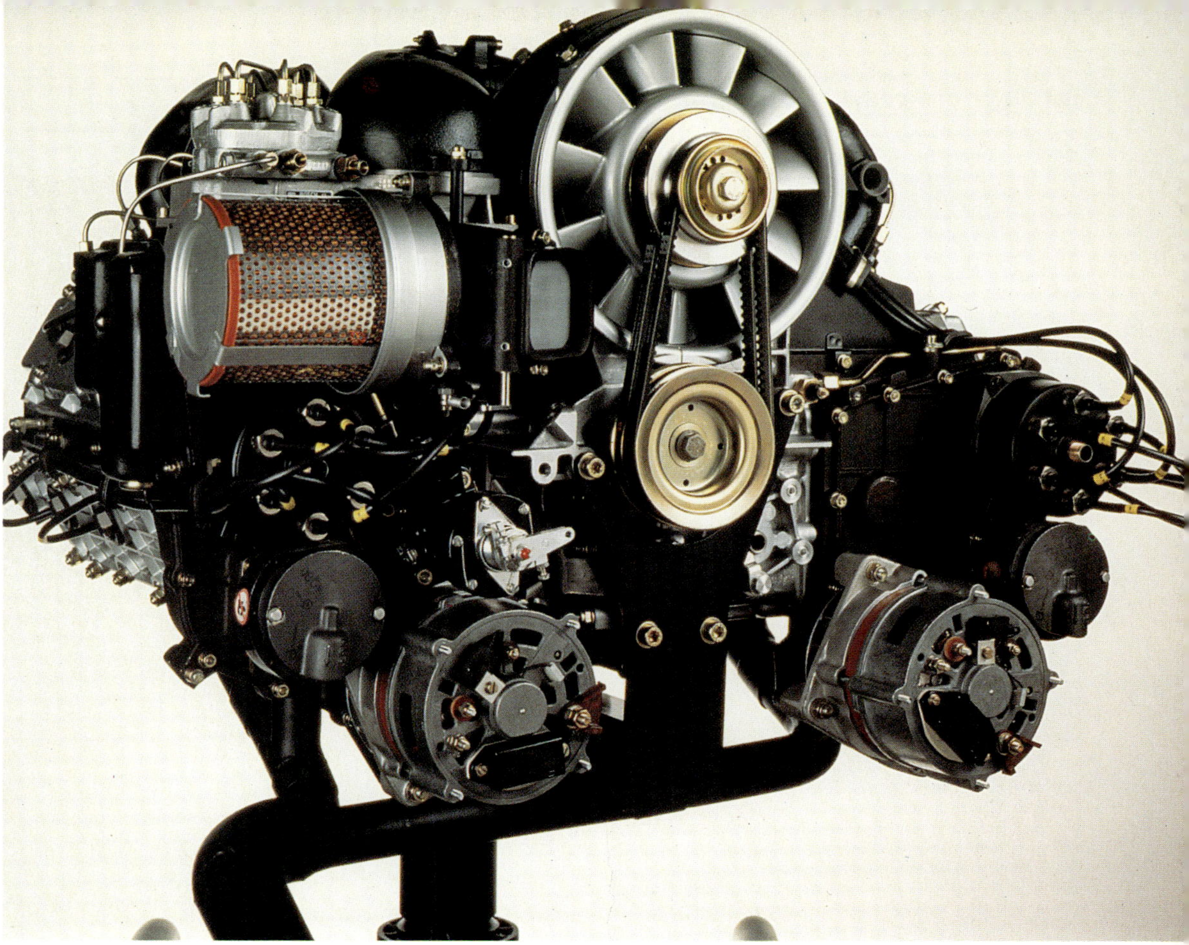

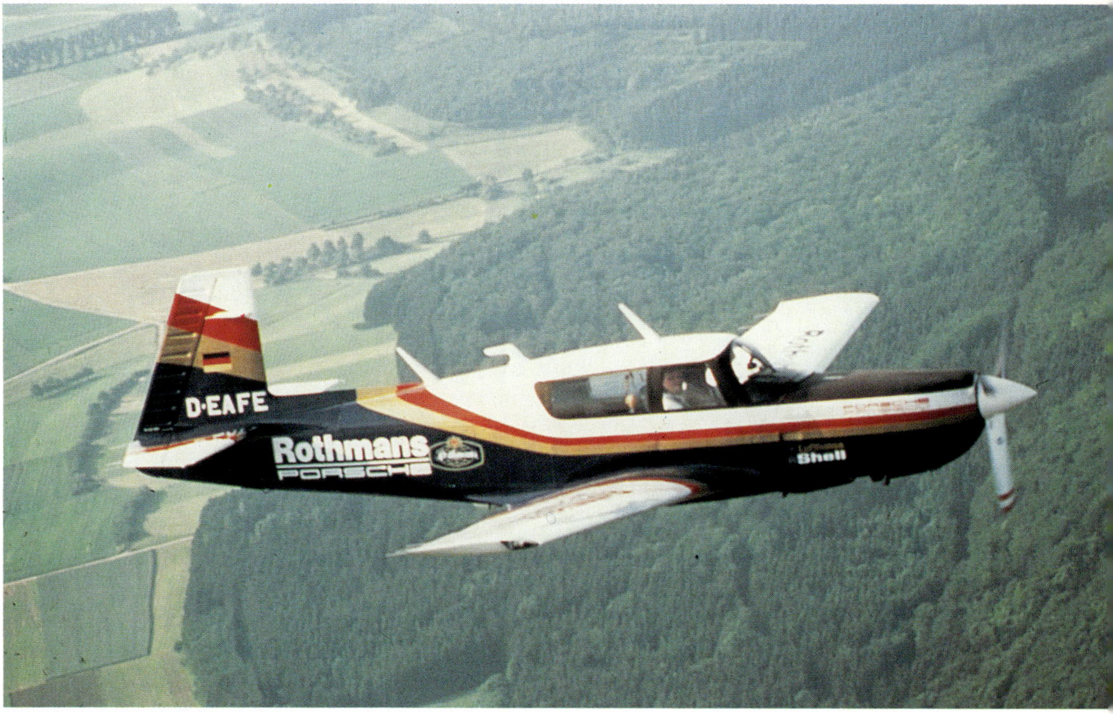
D-EAFE
Rothmans
PORSCHE
Shell

Porsche Longlife Auto of 1973. This design study used only 30% of non-recyclable material (70% is the norm) and demonstrated minimum ecological damage design principles.

However, Weissach's primary role is to ensure that the company remains ahead in the technological race and continues to produce a range of superb cars that give their owners "driving in its purest form". This philosophy is coupled with a constant and meticulous attention to production quality ensuring that all the current Porsche cars are truly "cars that every engineer wants to drive, yet those that any young girl could drive easily", and some of today's 'young girls' drive and race them superbly well. Exclusive and expensive they certainly are but automobile exotica they are not; Porsches are fast, safe and reliable. They are above all, practical supercars. Your Porsche is the result of extensive research and development and the product of a company with an unrivalled record of achievement in engineering design, innovation and development.

Whether this is your first Porsche or, as with two-thirds of owners, a replacement, you want to get the most from your car. Your choice of car says a great deal about you but the manner and skill with which you drive it, will say even more. We hope that this book will help you to enjoy the superb performance of your motorcar, in safety and with consideration for other road users.

Why another driving book? Because although improvements in roads and motorcars have been considerable, some 90% of accidents are still attributable to driver error. Moreover, as the owner of a Porsche, we believe that a driver's book devoted to your marque is the very best way to help you match your skill and expertise to your car's excellent handling and performance. Fast, safe and considerate motoring is essential to driving pleasure; and safe driving is our ultimate and overriding aim. This book is for those who take a pride in excellence and demand from themselves excellence in their driving.

This book is unlike any other driving book you may have read. In the first part of the book we describe the mental, physical and technical principles that are the foundations of a driving philosophy and how to put them into practice in your Porsche. These techniques are designed to help you to develop a smooth, swift and safe driving style so that you will obtain the maximum pleasure from your car; safely working towards the goal of matching your driving skills to your car's potential. Although written for you, the Porsche owner, the chapters on advanced driving techniques describe the best methods of advanced driving that the excellent motorist should employ in any car. The authors' aim is to help you match your skills to your car's performance.

The second part of the book analyses the particular handling characteristics of Porsche models, and deals exclusively with the necessary skills of handling and car control applicable to your motorcar. Together we shall examine the handling characteristics of your Porsche and explain how they can be modified by changing the basic specification of your car by fitting optional equipment. These chapters, and those in the third section of this book, have all been written with the expert assistance of Porsche's test drivers who have had decades of experience testing and driving Porsche prototypes and models on roads and tracks all over the world.

Their continuous programme of test driving is but a part of the never ending quest for product quality at Porsche. Germany's smallest car manufacturer – and the "finest" as is often added when speaking of Porsche – has an obsession with quality control and a development programme "more in the direction of refinement than greater numbers".

Professor Dr. Ferry Porsche once described the 911 thus,

> "The only automobile you can compete in the East
> African Safari or at Le Mans, drive to the theatre
> and finally in New York city traffic".

The renowned versatility and flexibility of the world's finest supercar has been achieved by an unrivalled programme of research and development and a company wide obsession with product quality. We take you behind the scenes at Stuttgart and Weissach and describe how these programmes contribute to your motorcar's exceptional quality, handling and performance. Because it is in the factory and at Weissach that Porsche ensures that every model upholds the tradition of giving you, the driver,

"Driving in its purest form".

CHAPTER 1

ADVANCED USE OF THE BASIC CONTROLS

We start our journey through the art and science of advanced high performance road driving by considering the proper use of a car's basic controls. A modern motorcar is a very sophisticated and refined instrument, and no problem to anyone, locked in a garage. It is only when that someone behind the wheel fails to use the car's control system properly, and without a sense of responsibility, that the danger potential is realised. In spite of extensive research and development programmes, tremendous advances in handling and performance, and numerous new safety features it is still human error that creates 90% of all road accidents.

The human element in the partnership between man and machine is usually the 'random' factor, which is why skill, discipline, training and many years of experience, particularly 'learned' experience, are required before you become a master driver. Knowledge, enthusiasm, 'natural flair' are all essential ingredients, but to be a fast, safe driver you have to discipline yourself to avoid the over confidence and lack of responsibility that are precursors to disaster; "The accident waiting for something to happen." So how should you use the major control system in a motorcar, yourself?

YOU – THE DRIVER

The expert driver will not only be able to judge what is safe or unsafe under any circumstances, but what is good or bad practice. 'Good judgement' is a combination of many factors including seeing the particular circumstances and recognising the detail, knowledge of your own and your car's ability, anticipation conditioned by experience, smoothness of control function, and a disciplined sense of speed, safe speed, for the conditions. Your own attitude to driving swiftly and safely is vital and with many years of practice to try to attain perfection, you will find that the right mental approach to safe, efficient driving is created by consistent self-discipline, of the kind that comes from within rather than the restraint imposed by speed limits and rules of the road. If you lose your temper, for whatever reason, you lose discipline and may eventually create an accident. The first principles should always be total consideration for others and control of yourself, since you are the ultimate controller of your motorcar.

Any first class sportsman or woman will tell you that discipline, skill, enthusiasm, dedication and concentration are minimum requirements for success. To be a very good as opposed to an average driver the requirements are the same, and concentration is vital. The driver capable of intense concentration will be able to see small detail, assess its value and foresee what can reasonably be considered the most likely to occur. This almost intuitive ability, which in the most expert driver is developed to an uncanny degree appearing almost psychic, is invariably lacking in the less experienced. Continuous application of self-discipline produces the ability to concentrate sufficiently to drive a motorcar with complete mastery at all times. Although most people can only concentrate for very short periods, you can improve your span of concentration by practice. The more you concentrate on your driving to the exclusion of everything else, the better your concentration becomes; and the more responsible a driver you will be.

Self-discipline must be matched with self-criticism, perfection does exist but no one is perfect all the time. Practice, particularly well directed practice, will create a good driver, but first take training of the right kind as expert advice is a great help (two brains are better than one). Listen to your mentor, watch, study and learn. When you drive, pick out the good drivers, watch them and you will see that the expert is swift and sure, safe and sound. The master driver shows by example that he or she is an expert by their consideration towards others, using their experience of driving not to 'pull a fast one', but to help and assist other road users to drive safely with a 'remote' following distance and sense of speed that gives everyone time to react. They are high achievers, people who are never satisfied with their own performance and always searching for perfection.

Recognising the difference between good and bad driving requires anticipation of what others think is good and bad, which is only developed with much experience of the behaviour of other road users – a vital part of road sense. The expert will 'blend and fit' his or her driving into the general pace, naturally and deliberately, ensuring that others always know what their intentions are. Although they will probably drive differently, in say Italy or France, in order to 'blend with the flow' they never let this consideration overcome their judgement of what is safe and sound.

The quality of our motoring is determined by the way we think about it. The best drivers often possess extreme intelligence; not always, but often! The crucial factor is the *concentrated application* of an intelligent and enquiring mind, and an ability to drive skilfully and calmly under all conditions, including the pressures of competition racing or rallying. Then you may develop, eventually, an almost bottomless pit of skill which can be held in reserve.

All of these qualities are admirable and necessary, but somehow they are not enough to produce a safe, fast driver on the highway. You must have anticipation conditioned by the right kind of experience; experience in the proper use of speed, and a sense of the safe speed for all conditions that enables you to travel safely, more relaxed at a higher speed than the majority. Always keep a vast safety margin in reserve. Develop your sense of 'pace' to a journey so that you can estimate your time of arrival; with experience you should be able to judge it to within minutes. Twenty five years experience of the High Performance Course has shown that very good road drivers are the most unobtrusive, so good they move quickly across the countryside, hardly noticed. To the inexperienced, they rapidly disappear into the distance; or they pass unseen!

Intangible assets of mind and body sometimes produce great 'flair' that are part of a person's make up at the start. This 'flair' to handle the car well produces a smooth, effortless and 'flowing' style which looks deceptively easy and is in complete sympathy with the car. This great skill and self discipline can transform driving into a fine art, impossible for the less able to copy. However, with the right mental attitude and physical qualities, almost any reasonably fit person can achieve a safe and swift standard of motoring.

Inexperienced drivers will often drive fast, through ignorance as well as over-confidence, and rely on their reactions to keep them out of trouble; they frighten their passengers and make violent use of the controls. Their poor anticipation and planning causes them to stamp on the brakes, grab and jerk the steering, unbalancing the car in the process. Their rear seat passengers often find it difficult to avoid bouncing off each other! The master will make you feel secure with no moments of indecision, harsh braking or steering. Progress is absolutely safe and just one long 'fluid flow' of motion which neither jars nor distresses. There is a contented, pleasant, 'social' atmosphere in the car. The driver is never heard to complain about the bad behaviour of others, all situations seem reasonable and expected, they 'open up' as the master arrives. Very rapid progress appears to be natural, swift and smooth, flowing with high road speed in comparison with the stolid, pedantic plodding of the inexperienced driver who invariably complains about others – drives too fast in town and too slowly in the country.

The physical frame within which your thinking, seeing, hearing and motor senses work is, of course, important. If you are going racing you need to be very fit and capable of strenuous physical exercise. As in any sport you have to be fit enough to sustain a sufficient level of concentration to maintain top performance throughout the stress of competition. Perfect vision, hearing and physique *all* have an influence on concentration which must be sustained throughout your driving. The physical stress of racing is extreme, Grand Prix drivers are Olympic standard athletes and heavy saloon and sports racing car drivers need to be. Whilst it is not necessary to be all that fit to be a swift, safe motorist, you need to enjoy good health – walking, gardening, or the occasional visit to the shops on a bike (instead of the car) may be sufficient. For the brain to be efficient, it needs a healthy body and plenty of rest – start driving already tired and not on top form mentally and physically and you will carry a handicap that can cost you, and other road users, dear.

If you have any doubts about your vision, and in our view your vision should be regularly tested (pilots have a complete 'physical' every 12 months), go to a reputable optician for an eye test. If necessary invest in the appropriate lenses and comfortable frames; comfort is very much more important than looks! Remember that peripheral vision is critically important and some frames limit your field of vision. Always ask your optician for his or her advice. Alternatively, consider the excellent contact lenses that are available. They are exceptionally good for some people and are often a preferable answer to glasses; but take your optician's advice. Glasses or contact lenses (if you need them *always* use them) are your life line to the action around you and as you will wear them for long periods make certain they are comfortable. Both need cleaning regularly and contact lenses must be cared for according to the instructions. Dirty glasses or lenses limit your vision as much as dirty headlights; both should be avoided. There are some motorists driving today who would not currently pass the Ministry of Transport test of vision; remember their limitations but never forget your own. Be particularly careful about the use of sunglasses when driving, especially in conditions of extreme light and shade. Where the sun is high and strong, good quality sunglasses can be invaluable in protecting the eyes and improving vision. The crucial factor is your vision, and glasses with simply 'coloured' pieces of glass instead of good lenses can be disastrous; for you and for other road users. As with prescription lenses, quality is very important, as good quality glasses will reduce glare but not vision.

Your Cockpit Drill

Fatigue is a function of seat position and comfort at the 'wheel'. Porsche seats are adjustable, at least, in terms of back rake and squab position. Some models have electrically adjustable seats that have seat back, rake and lumbar adjustment with squab height, slope and position. Use all the adjustments to ensure that your seat is exactly right, placing you in an optimum position in relation to all the controls. First adjust the position of the seat squab so that you can fully depress the throttle pedal with the back of your thigh in contact with the squab edge. If you have a squab height and rake adjustment raise the seat until your eyes are level with the centre of the windscreen. The front of the seat squab should be adjusted for rake so that it supports your thighs just behind the knees. When the seat squab is in

Check your seat position by running each hand from the bottom to the top of the steering wheel; left hand clockwise, right hand anti-clockwise. With your hands at the top of the wheel your spine should be firmly against the seat back.

the correct position, adjust the rake of the back so that you can reach the top of the steering wheel at twelve o'clock with both hands. Press the small of your back firmly against the seat by bracing your feet against the foot rest to the left of the clutch pedal and against the bulkhead to the right of the throttle pedal. Move your hands around the rim of the steering wheel and ensure that both can comfortably reach the top and put firm pressure on it. The base of your spine must be firmly against the seat back. You will get the best support when your legs form the sides of an equilateral triangle with the spine as the apex and the feet as the extremities of the base. The right ankle should be at an easy relaxed angle, not bent tight back against the joint; a flexible ankle can react quickly in an emergency. The muscle at the top of the lower leg should not be stressed, holding off the throttle, as it is when you sit too close. You must be able to reach the gear lever in its furthest position without stretching forward and to fully depress clutch, brake and throttle pedals without stretching forward.

When you are sitting comfortably, run your eyes over all the controls and check that none of the instruments are obscured by your hands when they are in the quarter to three position. Be certain that you can operate all the stalk controls for signalling, screen wipe/wash, headlight dip/full beam/flash, without removing your hands from the steering wheel rim. In a strange car be absolutely certain that you know where the controls are before you set off. Fortunately, most cars now have a standard layout for stalk controls. Check the

Use the foot rest to the left of the clutch pedal when not changing gear.

position and operation of all the other controls so that you can operate them without taking your eyes off the road ahead.

Next adjust the rear view and external mirrors. This should *always* be done on flat ground so that you can be certain to have the maximum field of vision from all three mirrors. With near and offside external mirrors (the latter is compulsory in some countries) and the internal rear view mirror properly adjusted you should have a complete rear view. Porsche external mirrors are adjusted by a 'joystick' on the driver's door just forward of the window control. A two position switch on the central console or under the instrument panel (depending on the model) determines which mirror you are adjusting.

Your passenger should now see you sitting comfortably, well back with your arms bent (not straight like a 'boy racer') and with a light touch on the steering wheel. You should be relaxed and totally familiar with all the controls and, having finally checked all the instruments, you can fasten your seat belt and make sure the lap and diagonal straps are free from any obstruction and have not caught any of your clothes. Check that all the doors are secure and if you have a passenger that their seat belt is correctly fastened.

Familiarity with a motorcar can, as in other situations, breed contempt, but for the experienced and interested motorist it produces a keen sense of awareness and relaxation coupled with a greater sense of knowledge as to the car's limitations. 'Polish' is developed from a complete understanding of what the control functions do to the mechanical components in the car and how to operate them. Experi-

(Above and opposite) A relaxed seating position with arms slightly bent will give you maximum control and comfort. Comfort is important on all journeys as it aids concentration.

enced drivers are adaptable, settle very easily to a strange vehicle and make no fuss about an unusual control layout. They just take a mental note in a thorough cockpit check and get on with driving; making it look easy and natural. All cars are different and require differing techniques, but difficulty with the controls of a strange car is potentially hazardous. The greatest asset in adapting quickly to a strange vehicle is to have the correct 'pattern' of thought and action, coupled with sound technique; many drivers do not . As a guide, the following is a check list which you should always use:

1. Check all baggage is stowed *and secure*.
2. Check lights; *all of them*.
3. Check that doors are secure.
4. Adjust your seat as described.
5. Fasten seat belts (passenger(s) as well!).
6. Adjust internal and external mirrors.
7. Check handbrake is on.
8. Check gear lever is in neutral.
9. Start engine.
10. Check oil pressure, temperature and that all warning lights are out (oil, ignition, etc.).
11. Conduct a static brake test.

Even if you are fortunate and have the flair to drive well you will need training, not only to instil the correct mental attitude but also because the natural method of handling and coordinating control

functions the easy way will not be correct; you may put the clutch down too early before you reach for the gear lever, cross your arms when steering and so on. There is no substitute for expert professional instruction from the start, ideally by a police trained instructor who teaches the principles of roadcraft and understands the importance of applying road discipline. You will then develop the correct discipline of thought, action and movement that is systematic and has reason and logic. Once formed it stays for the rest of your driving life; good instruction is important from the start.

(Opposite above and below) Stow everything securely; the distraction of 'objects' rattling around inside the car can be dangerous. It is surprising how much luggage can be securely stored.

The physical movements of good driving are often slow, almost lethargic, because once the direction of effort is found minimum effort is required, and if thought processes are early and quick, early timing can be applied without unsympathetic sharpness. Even when driving fast, control movements should be in complete sympathy with the car and should 'flow' smoothly from one control to the next, blending and coordinating one control function into another. The impression given to a passenger is that driving is automatic, responding to situations without thought although nothing could be further from the truth. Every aspect of driving is a problem and is only solved by thought processes and continuous concentration with early application of technique that is developed by fine timing and perfected by years of self-criticism.

Before you can develop the advanced skills of the master road driver, or those required for successful and safe participation in hill climbing or racing, it is essential to understand the way the controls are linked. Think of the functions of the driving controls in pairs. It may be obvious to you to link the gear lever with the clutch but they must also be used in the correct sequence. We all link firm depression of the accelerator with successfully rising gear ratios to increase speed rapidly, but do you think of car sympathy and use of the rev counter? Remember that the accelerator works both ways and appreciate the value of deceleration as well as acceleration sense. Then link firm acceleration with steering to 'tip and tilt' the car in 'balance' round a curve and appreciate the build-up of side forces created by increasing speed in a bend. Steering is never improved by harsh braking; as you approach a corner link braking and steering in your mind, by always having both hands on the wheel when the right foot is braking, only move from that for very fast or competitive driving when pedals are close for easy and natural 'heel and toe' operation. Next, link your driving mirror(s) with braking by teaching yourself to look before braking and take prompt, correct action upon what you see! Always link your driving mirror(s) with signals by giving yourself reason to signal. Time each signal for every situation, after assessing the value of what you see in the mirror.

This 'thinking in pairs' is crucial to an understanding of the proper use of controls and for understanding the way forces act upon a car. So read this next section carefully, it will be invaluable in your future enjoyment of driving.

THE CLUTCH

The clutch is perhaps one of the most abused, and yet important, controls in a manual motorcar. Many more clutches are damaged or suffer a rather short life span because of improper use by the driver, than through mechanical deficiency. A clutch is a make or break between your engine and the transmission; it can easily be mis-used. Expert use of the clutch is essential to smooth, flowing driving which is more comfortable for you, your passenger(s), and will enhance the mechanical life of your car.

Don't start off by running the engine too fast for work effort needed to move the car from rest. In order to achieve a smooth gentle take off there should be 'silky' coordination between the handbrake, power and clutch, to glide away at no more than twice the idling speed of the engine on the level. Only when the clutch is fully 'home' or engaged is the power used firmly to accelerate away; what is lost by engaging the clutch sympathetically you should make up by using the power and acceleration of a fast car. Treat the clutch gently and never ride the thrust race at traffic lights to be away first, crossing the line on amber, committing an offence. Better to stop half a car length back from the line as a margin of safety for pedestrians and to gain a better view of the junction. Then 'roll' away safely and gently as the traffic moves off.

Once on the move, the clutch should be treated exactly like a 'make or break' electrical switch – either in or out, on or off. It should never be slipped or dragged and smoothness should depend on your skill in controlling the engine speed with the throttle. For example, it is bad driving to try and make a very sharp left turn in second gear, slipping the clutch to increase speed. In the correct gear, first, no slipping of the clutch is required. There is never any need to coast or slip the clutch or select neutral when the car is in motion. It is a potentially dangerous procedure, particularly when descending a hill or approaching a crossroads. It is dangerous to rush up to a hazard leaving your gear change too late, or coasting near to the hazard (particularly if you miss that rushed change), with disastrous consequences for you and the traffic you may emerge into. Better and safer to always be under control in the right gear, at the right time, in the right place.

Unnecessary changes, particularly changes that use the clutch to reduce speed, impose wear and tear on the whole transmission system. Think of ten such unnecessary changes in five miles, then multiply that by your annual mileage! Brakes are designed to reduce speed, so use them. The only occasion when it is necessary to use the engine to reduce speed is going downhill. Then, reduce speed by braking at the top of the hill and use the appropriate gear to maintain that speed; but braking where necessary.

Always avoid the clutch damaging habit of pushing the gear lever through the synchromesh at too high a speed for the gear, then leaving the clutch dragging against an idling engine, with accompanying secondary braking, causing more clutch wear; sometimes this will happen twice at the same hazard! Some drivers will persist with the habit for years, even enthusiasts competing in races or hill climbs. Watch how such drivers will charge up to a corner, brake long and hard, transferring the weight to the front wheels, then make a rough, jerky change down, letting the clutch in harshly and banging it against the idling flywheel of a high compression racing engine. The lightly laden back wheels will lock-up as the driver turns into the bend. Sad that such drivers should spend hard earned money improving their car without first developing their driving skill.

The Gear Lever

If the gear lever and clutch are not used in the correct sequence by early preparation, gears can easily be missed, particularly when driving a strange vehicle. The quality of gear changing will make or mar your driving and superb technique is the hallmark of a master driver. A large proportion of bad gear changes are the result of poor understanding of the forces acting upon the clutch and gearbox. Do not forget that the synchromesh is basically a friction device and for long life it is preferable to use it as little as possible. The use of double-declutching is quite important, and it is necessary to understand why.

In many high performance cars the gearbox is somewhat heavy to engage.This is because gearboxes in such cars have to transmit a great deal of engine torque and power, so the gears tend to be larger. This results in greater inertia (or resistance to change in velocity) of the primary and lay shafts at high engine revolutions. An unsympathetic driver will have a tendency to hurry the gear change (often evidence of poor anticipation and advance observation), thus pulling the lever hard against the resistance of the synchromesh, and raking the synchro cones. Modern gearbox design and materials have undoubtedly improved the life and toughness of transmission units and they will often stand this abuse for many years, giving a lazy driver the excuse to say "double-declutching is unnecessary". Nevertheless, the smoothness of a 'knife through butter' gear shift at high engine and gearbox shaft speed is better for both car and passengers, if accurately accomplished.

In cars with the clutch behind the front mounted engine, and the gearbox and drive unit in the rear, the synchromesh has to cope with with the inertia of the propeller and lay shafts, making double-declutching even more beneficial at high engine speed. With small, light gearboxes, it is less necessary at low road and engine speed but the skill to do it if necessary is essential in any good driver. You can then drive without the clutch if it fails to disengage (fortunately a rare occurrence) and proceed to a garage. Changing gear should be

(Above and opposite) Palm the gear lever, don't 'grab' it. The gear lever (like the clutch) is there for a purpose, not as a hand rest.

2:26
Panasonic

a smooth, easy transmission of engine power, not a hasty, jerky, uncomfortable and fundamentally destructive action. Let us run through the procedure of changing gear, as it should be done.

TO CHANGE UP, prepare by 'palming' the lever for direction of effort and cover the clutch with the left foot. Then, with the clutch down and the power off, select neutral and with the clutch up, time the fall in engine revolutions so that the engine speed is correct for the next higher gear. Depress the clutch and complete the change with a flowing, synchronised movement of the clutch and gear lever.

TO CHANGE DOWN, palm the gear lever and cover the clutch pedal with the left foot. With the clutch down and the power off, select neutral. Then, whilst listening to the engine with the clutch up momentarily, increase the engine speed revolutions by controlled use of the throttle. Next, with engine revolutions sustained, depress the clutch, select the gear and release the clutch. When changing down, always match engine speed with gear and road speed so that the clutch engages cleanly without any 'drag' from the idling engine, the unit being brought up to the correct engine speed. If you are inexperienced, to complete a change that is smooth and not rough, do not 'blip' the throttle. When a fast downshift is required, use this technique only if you have the skill to match engine speed precisely by blipping the throttle. Try for smoothness not speed; the latter will come later.

The slower and more precise way is to develop your talent by listening to the engine of your car, getting to know its sound at different engine revolutions and practise on say an empty dual carriageway until your gear change is smooth, unobtrusive and 'clean'. Precise coordination of the double left foot movement of the clutch pedal and related gear lever movement in a double-declutch change is not so important as assessment of exact engine speed in relation to car and road speed. More harm will be done to the synchromesh by double-declutching if the engine is not perfectly matched to car and road speed. On a clear road glance at the rev counter while you change up, the needle should fall and then stay still (not fall too far then fluctuate back to the correct reading). When changing down the rev counter must rise to the correct engine speed, with the revs kept steady while the change is completed. Swiftness of changes will come with practice – so never sacrifice accuracy and smoothness for speed.

It is better to flex the ankle as well as knee joints and only dab the clutch over the 'point' of clutch control to free the drive – it is less necessary to fully depress the clutch when double-declutching than when using the synchromesh to match gearbox speeds. The more the clutch pedal is depressed, the more the 'fingers' of the diaphragm clutch spring flex – an analogy is to bend your own fingers back and see the stress in your hand. It does not matter whether the clutch plates are far apart or only just clear of each other, as long as the drive between the engine and the gearbox is freed by the disengagement. Clutch control is improved by leg flexibility, the heel in contact with and sliding along the floorboard of the car. Many instructors fail to advise novices to keep heel/floor contact as a position for a hill start with the consequent lack of sensitive control. The back of your thigh should be in gentle contact with the seat cushion for support in cornering and long periods at the wheel, not 'cramped up' with the knee bend that comes from a seating position too close to the pedals, resulting in 'bottoming' the clutch unnecessarily.

BRAKING

There is more to learn about braking than any other control function and many drivers do not appreciate how far it takes to stop a car in an emergency. As a general guide to braking distances on a good, dry, tarmacadam road surface you square the speed in miles per hour and divide by twenty, then add the driver's thinking distance. A superbly fit professional racing driver takes approximately 0.2 seconds to react and apply the brakes, the average driver perhaps two or three times that. Therefore, at 60 mph (= 88 ft/second or 26.5 metres/second) you will travel a minimum of 35 feet *before you commence braking*. The table below gives thinking and braking distances in feet and metres.

APPROXIMATE STOPPING DISTANCES IN PERFECT CONDITIONS

SPEEDS		STOPPING TIMES AND DISTANCE							
			Thinking Distance			*Braking Distance*		*Totals*	
MPH	KM/H	Time (secs)	Feet	Metres	Time (secs)	Feet	Metres	Feet	Metres
20	32.18	0.5	14.66	4.47	1.36	20	6.10	34.66	10.70
30	48.27	0.5	22	6.61	2.04	45	13.725	67.0	20.605
40	64.36	0.5	29.32	8.94	2.73	80	24.40	109.32	33.34
60	96.54	0.5	44	13.22	4.09	180	54.90	224	68.32
80	128.72	0.5	58.64	17.88	5.45	320	97.60	378.64	114.48

Even at 30 mph/48.27 km/h you will travel 22 feet/6.6 metres before your car starts to slow and in perfect conditions, a further 45 feet/13.7 metres before you stop. Drive through the built up area you are most familiar with and judge how far these distances are on the road. A car can be a lethal weapon! Table 1 is based upon a thinking/reaction time of 0.5 seconds, but we know that unless you are instantly alert to a danger and concentrating exclusively on your driving, your reaction times will be even longer. Concentration and mental application to the exclusion of anything irrelevant will improve your reaction time and may save your life, or someone else's!

It takes an astonishing amount of energy to decelerate a moving vehicle. In fact it takes exactly the same amount of energy to decelerate from one speed to another as to accelerate between the two speeds

– the forces of inertia are always against us. The actual energy required for braking is a function of the speed and weight of the car;

$$\text{Energy (Lbs/Ft)} = 0.0335 \times [(\text{mph max})^2\ (\text{mph min})^2] \times \text{gross weight}$$

A rough calculation of the braking energy put out by a Porsche 917 over a twelve-hour endurance race would bear comparison with the energy needed to supply the electricity of a fair sized town!

Braking distances are obviously a function of the rolling resistance of the tyres (not much), aerodynamic drag (only at over 60 mph), friction in the entire mechanism (not much) and mostly the braking system which converts the vehicle's kinetic energy into thermal energy; which is why brakes get hot. But tyres have to grip or the car's brakes will not stop the car. Therefore, always consider the reduction in adhesion generated by different conditions. Assume ideal conditions and an excellent road surface as used on the approach to pedestrian crossings and take that grip value as 100%. Ice will provide 5%-10%, so you will just travel on. Even good dry tarmacadam will only give 80%, and wet tarmac will give 50%. A great deal of difference.

THE FIRST PRINCIPLE of braking is acute observation of the road surface. Always relate conditions to the grip provided by your tyres. Many drivers fail to 'read' the road surface at all and never notice the critical changes. Rally drivers are most experienced at the assessment and they have fantastic judgement. It has almost become instinctive for them, based upon years of driving on ice and snow, gravel and tarmac and every combination in between. They know the best choice of tyres and suspension settings to give the best handling and grip.

THE SECOND PRINCIPLE is that you must match your braking distance to your depth of vision or sight line; can you stop in the distance you can see? This 'sight line' must be of the whole road surface, allowing no 'dead' ground. Precise judgement of speed and distance is essential for safe driving. Remember it is not the car's capacity to corner that matters in a blind bend or crest, but your ability to stop comfortably and well within the distance (with a safety margin) that counts. Realising that the surface is bad or the distance to a hazard too short when you are upon it, is too late! The road ahead must be read in advance and speed adjusted with ability to stop.

Most road drivers never experience the limit of their motorcar's braking capability until an emergency happens; then they discover that the brakes pull left or right, or lock up front and back; which may spin the car. Emergency braking is always hazardous because of the effect braking forces have on the car. Under braking the car's weight is thrown forwards, in a straight line along the car's longitudinal axis, forcing the front wheels into the road and un-weighting the rear wheels. Don't brake over bad bumps or potholes, let the suspension support the car normally, time the rebound of the suspension to make light of the shock. Always keep both hands on the steering wheel to keep the car straight and true; and always try to brake in a straight line.

THIS IS THE THIRD PRINCIPLE OF BRAKING, always try to brake in a straight line, whenever possible. Braking on a corner reduces the adhesion between tyres and road under any conditions, as the tyre is immediately subjected to two forces, one braking and the other cornering. This places great strain on the tyre and can easily cause breakdown of adhesion; it always reduces the cornering ability of the tyre. Watch carefully how some drivers in a motor race will arrive too fast at a corner, turn the front wheels whilst still braking heavily and occasionally fall off the track! Braking should always be tapered off gradually before cornering, otherwise the tip and tilt of the car under the combination of retarding and steering forces will cause dramatic instability. 'Trailbraking' into a bend is only for the very experienced driver and never to be attempted by novices.

The importance of keeping both hands on the steering wheel under braking for a hazard, must not escape you. Most cars have a tendency to be unstable under braking and control of steering is therefore critical. Even if the car is stable potholes, changes in road surface and other factors will influence the car's passage. Release the brake only when your road speed is correct for the conditions ahead of you, so that you have plenty of time to make the gear change separately and your right foot is free to synchronise engine speed. Remember that gear changing takes time, at sixty miles per hour if your gear change takes two seconds, you will travel 176 feet. So do not brake, change down, then start braking again for the same hazard. Be ahead of the car and anticipate, in the first instance, the precise speed for safe negotiation of the hazard. Much better to be a little too slow and make the same gear change covering less distance.

Your continuous and correct judgement of the safe road speed for the conditions is the key to safe, fast and smooth motoring. It is speed and car control that demands which gear you will require. Until you can judge correctly the speed you will require, how can you possibly choose the correct gear?

The technique of braking steadily in a straight line (keeping both hands on the wheel while the car is unstable because of the transfer of weight to the front wheels), easing off the brakes and then changing 'cleanly' into the appropriate gear for the speed, is to be recommended for gentle or moderately firm braking in road driving. If you imagine that your braking pressure should be as in the diagram overleaf, you will find your braking smooth and avoid that final 'jerk' of the inexperienced driver.

THE FOURTH PRINCIPLE of braking is progression, the firmest application is steady and in the middle of the braking distance. Establish 'contact' with the brakes with a light and delicate, initial pressure of the pedal, then increase pressure progressively to the steady application required and gradually taper the braking pressure towards the end of the braking period. Faulty assessment of hazard potential, appropriate speed and distance will make you increase the pressure at the end of the braking period; or you will suddenly release the brake hurriedly as though the pedal was 'hot', to transfer the right foot to match engine speed to change down. Remember that secondary braking for a hazard indicates a lack of judgement and deliberation.

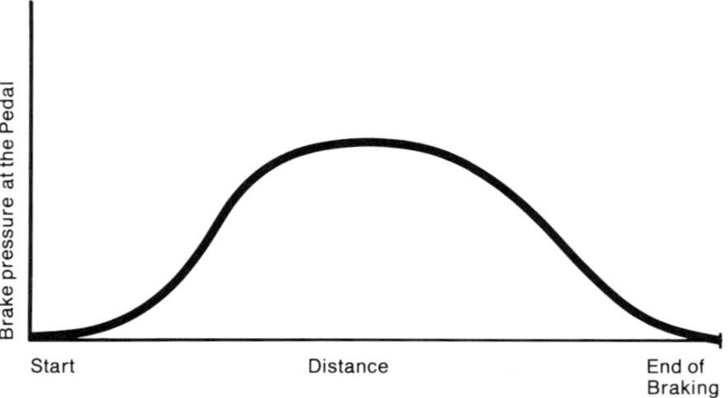

Apply the brake pressure 'progressively'. Ease it on and ease it off.

Practise braking from high speed with light, initial pressure just 'brushing' the brakes, then progressively transfer weight to the front by firm braking, and from then on keep a steady firm pressure if required, progressively easing off as you reach the correct speed before the hazard. Loss of speed should be constant and steady. When braking to a halt, remember that stationary friction has a greater effect than 'running' friction, so almost all pressure must be released as the wheels come to rest – without a jerk or sudden stop. Why is it that so many drivers jerk their passengers with a sickening lurch just as the car is stopping? Even their best friends will not tell them! They keep the same pressure on the brake when stopping and the brakes have a binding effect just as the wheels come to rest. It is best to leave more distance to stop and release sufficient pressure to avoid that head snapping jerk. When braking or cornering, never place the car very close to the edge of the road. The gulley is often a repository of stones (marbles) and grit that will dramatically reduce tyre grip. Keep one and a half feet as a margin of safety in case harsh braking or road camber draw you off line.

The technique known as 'heel and toe' is not taught at Police Driving Schools, nor was it taught at the Jim Russell Racing Drivers School when John Lyon was an instructor there. Jim Russell's view was that like all things in life, it is better to learn something properly, one thing at a time. However, once you have completely mastered the essential judgement of speed and distance, the technique of raising engine speed with the side of the right foot, whilst braking firmly with the ball of the foot, can be an asset for fast driving on the road or circuit. However, you will only accomplish this technique safely in a car with the appropriate pedal layout, like your Porsche; never attempt the

technique in a strange car or when braking gently. The cardinal rule with this technique, which should always be observed, is that all braking should be executed with the ball of the foot firmly and squarely on the brake pedal. If your foot slips off the brake as you go into a hairpin or junction, you could hit an oncoming motorist!

'Heel and toe' technique requires a great deal of practice and expert, professional instruction.

Practise this technique only on a (relatively) empty race track or a clear empty road, using the following procedure. Reduce road speed by steady braking and as soon as your road speed is well within the comfortable engine revolutions of the gear you will select, disengage the clutch and select neutral, then lift your left foot, taking the pressure off the clutch, and roll your right foot (maintaining firm brake pressure with the ball of your foot and blipping the throttle with the side of your foot) to maintain or increase engine revolutions to match road speed for the next gear, and complete the double-declutching gear change. Continue to brake evenly with reducing pressure up to the 'turn in' to the bend. If you miss the gear change, you have a better chance of making a second attempt without being in a 'box full of neutrals' in to the corner. Also, your braking distance can be shortened and moved up to the corner with greater flexibility of judgement to adjust braking pressure. This technique gives greater security providing that you discipline yourself *not to always* use it for gentle road driving, but to keep the technique in reserve as something left if you are driving swiftly.

An occasion which calls for the greatest care is when the brakes have to be applied while cornering fast. In this case it is often the inside rear wheel which locks first and causes the rear end to 'break away', particularly with a front-wheel drive car. For a hump back bridge, brake to the crest throwing weight forward and slowing down. When you see safely over the top as you begin to drop, get on the power and throw the weight back, letting the suspension settle square down on the other side. It stops 'nose diving' and crashing the front suspension.

TAILGATING, to which we shall refer later, is in our view inexcusable and lethal. There is never any reason to intimidate the preceding driver to speed up or to move over, or wilfully prevent a driver from filling a gap. There are far too many fatal accidents on British motorways caused by this lack of good, safe driving which determines that you will follow at a safe distance. That safe distance remember, is the distance in which you can stop safely and smoothly; not when brought to a rapid halt by the boot of the car in front, with possibly fatal consequences. Quite simply, there is never any reason to follow too closely to another vehicle, it restricts your view and is *dangerous*.

A driver's skill is usually inversely proportional to the unnecessary braking for a given mileage. Poor drivers have an instinctive tendency to use the brakes when they are unsure of what is going to happen ahead. Don't brake because it gives you more confidence, brake because you need to. Learn and use the fundamental principles of the expert motorist. Long range anticipation, precise judgement of speed and distance, detailed observation, correct assessment of the value of what is seen, timely and polite warning of approach and appropriate speed will reduce the number of unnecessary applications of the brake. You will also drive more quickly, smoothly and safely!

Braking in an emergency is a whole new and often nasty experience. Use the driving plan and always match speed with vision and you should avoid emergencies; or most of them! Nevertheless, it will probably happen and when it does the natural reaction is to 'bang' the brake pedal hard and fast – the wheels 'lock up' and the car slides straight on, out of control as you cannot steer a skidding motorcar. On a good dry road you will stop quickly, and on snow or gravel steering deflection will form a 'wedge' to help to stop, but not to steer. With tyre adhesion gone it matters little what you do with the steering wheel, the car travels on its line of momentum (See Chapter 7). Release the brakes with an angle of steering applied and the instant wheels roll, adhesion is regained and the car will steer.

The best braking grip, on any surface, occurs *just* before the wheels lock and stop rotating. The only way you can gain maximum braking is to lock up all four wheels and, if you can, depress the clutch. Brake not with a 'bang' on the pedal but with a quick and increasing push. The moment that you feel a slight bump, as the wheels lock, release *just* enough pressure for them to rotate again; but don't release them fully, keep some pressure on. Lock up all four wheels and then relax until they rotate – increase, relax, increase, relax, until the car 'stops'. Known as cadence braking this technique is not that easy. The crucial point is that whilst the wheels are turning with 'near' maximum brake pressure applied you will have control of direction.

The car with an anti-lock braking system will achieve this cadence effect for you, although you are applying maximum steady pressure to the brake pedal with the clutch down. A life saving device in the worst situations. We will discuss the effect and practice of cadence braking and anti-lock braking systems in Chapter 7. Here, just note that it is yet another advanced use of the basic controls that requires constant practice to learn, and it can only be safely learnt on a skid pan or track.

Acceleration

Acceleration sense is another sign of a master driver and a very necessary skill. Do not slip the clutch too much when starting off, twice the idling speed of the engine should be the maximum, even for a brisk start, and the clutch should be fully 'home' before the power is turned on, after starting or changing up. It is best to pull away smoothly and sympathetically with silky coordination of throttle and clutch. Use only half the available power and engine speed in first gear, for sympathy of the car and passengers, then full power and revs in second and subsequent gears if necessary.

Drivers tend instinctively to brace themselves against their own actions, and are not always aware of the discomfort from harsh acceleration or braking. Therefore, for the comfort of your passengers and less wear and tear on your car, apply power and brakes progressively for fast, smooth driving. You gain *nothing* in a car from harsh use of the controls; except excess wear and discomfort. Before and after each gear change the power is 'tapered off', and on respectively. Before the gear change, transition from acceleration (throttle depressed) and drive forward to overrun (shut down of the throttle), must be gradual and progressive. After the change open the throttle slowly, firmly, but progressively and to the *floor* when you require smooth, but firm acceleration. As an analogy, imagine twisting a piece of wire back and forth, it snaps with fatigue. That is the stress you put on gearbox and transmission by harsh acceleration, to say nothing of your passengers' necks.

The key is the ability to drive quickly, safely and smoothly, with varying speed for different road and traffic conditions with accurate use of the accelerator when braking is not required. The marked difference between the good, and the *very* good driver, is that for a given average speed, the latter will brake less. Follow the master along a twisting, winding country road and there is hardly a twinkle of brake lights to maintain a high average speed. Fine judgement and acceleration sense is closely allied to gearbox sense, and a driver without the latter seldom has the former.

Acceleration sense must be linked to a fine judgement of speed to give that essential ingredient in safe driving, *time to react* – not only for yourself, but for others to have time *to react to you.* If you are driving at the appropriate speed down a busy high street you may well find that others are pressing you. The majority of motorists will not be anticipating hazards, pedestrians, animals, or parked cars when they should be doing so. After all, in crowded traffic conditions there are more 'potential' hazards and expert drivers will match their speed accordingly. Remember your braking distances!

Always try to 'pace' your driving and arrive at situations as they are 'opening up' not 'closing down'. Apply acceleration sense to all facets of road driving, on approach to hazards seen well in advance, following others, or overtaking them. Use it with great skill to 'fit and blend' with traffic joining motorways and roundabouts. A powerful car with brakes and handling to match, is a great advantage for the ability to accelerate fast out of a zone of danger, or leave behind an inconsiderate driver who insists upon tailgate intimidation, following too closely for their own safety. Power is a positive means of creating space and, thank goodness, it is not illegal to use firm acceleration when it is clear and safe to do so. You will never become an advanced motorist unless you learn to use the full degree of acceleration; *safely*. How can you tell what the car is capable of if the accelerator travel is never taken beyond halfway?

The economic crisis in the 1970s had a marked influence upon driving behaviour, and even the best motorists have to be encouraged to use power as a tangible safety factor. At no time should driving technique to save fuel be detrimental to road safety; coasting down hills, keeping the same speed in the country as in the town, just to save time and petrol are wrong. Perhaps this is why many drivers seem to abhor the sound of an engine 'singing' at reasonable revolutions. This often causes them to try and negotiate a hazard in too high a gear for flexible control of road speed, with nothing in reserve – no power ready for the unexpected. Engine speed should always be chosen for safety and efficiency, controlling road speed is a primary objective. Drivers often change up too early, just because the unit makes a bit more noise. So motorcars make noise, so what? Isn't the sound of a Porsche 911 Carrera music to your ears?

Always vary speed with the changing road conditions using your judgement to match speed with vision, for safety. The first opportunity to overtake is never missed, if you acquire an instinctive assessment of your car's acceleration ability, and it is one reason why it is a good habit to use firm acceleration at every safe clear opportunity, if safe. Practise your use of acceleration, because acceleration sense is only acquired after many years of driving. A lack of safe overtaking skill is perhaps the greatest weakness in the majority of drivers in this country and is partly due to poor acceleration sense and a reluctance to use power when safe.

In overtaking the engine should be operating between maximum torque and power, fine judgement of road speed and appropriate gear being critical here. In most cars this will be from approximately 3000 to 6000 rpm. The accelerator should be depressed gradually and then fully to the floor. With a turbocharged engine the turbo has to spin at about 50,000 rpm at an engine speed of some 3000 rpm, so, as you depress the accelerator, glance at the turbo boost gauge and prepare for acceleration. Position, timing and appropriate speed are all critical, noise does not come into it. However, always change up 500 to 1000 rpm, below the red sector on the rev counter, for car sympathy. At these revolutions, inertia and wear and tear to the valve gear is extreme and there is no point in revving past the power peak which is usually 500 revs below maximum engine speed. When the engine noise sounds right, glance at the dial to check what sympathetic ears should be telling you. Get to know your car and its capabilities, instinctively, and drive within them.

On corners, increase the power with the expanding radius – but read the road surface. Only use full power early with a good dry surface. Most recognise this skill as acceleration sense when combined with steering, but the best drivers also have acceleration sense in a straight line and appreciate skilful use of the accelerator all the time they drive. The accelerator works *both ways* and the compression of the engine causes engine braking with the throttle shut, more so at moderate to high engine speeds – braking energy that is 'free of charge', as no petrol is used. Yet so few use deceleration with effect, most cruise in a high gear and use the brakes to slow. Only the best drivers have acceleration sense, but it is often latent and can be improved with demonstration, instruction and practice.

Gradients have a dramatic influence upon the choice of gear, i.e. low gear and high engine speed for steep hills, and a low gear is often necessary, even for gentle descents where no reduction in speed is required. The use of power for the hill should be planned ahead and the best speed and gear chosen well in advance. It is a poor driver who waits until the engine staggers and labours up the hill before a gear change is made. As soon as the road speed falls away on a steep hill and engine speed is appropriate, the correct gear should be selected. In this instance, have the engine just pulling before the change and match engine speed with gear and road speed to make a 'clean' change; double-declutch if required. Descending a steep hill, use the braking effect of the compression of the engine, select the same gear ratio to go down the hill as you would to go up, and take it by the sign saying 'low gear now' at the top of the hill, not dangerously half way down.

You should be capable of missing out intermediate gears. It is rarely necessary to go 'through the box' on the way down. Changing up a 'one to three' or 'two to four' change can be used, if *firm* acceleration is required away from the hazard, for a safe gap behind, then the high gear for cruising at the speed limit.

Excellent and skilful acceleration sense makes overtaking a line of vehicles a well timed, brisk and considerate manoeuvre, achieved without touching the brakes and using the acceleration of a powerful car for the absolute minimum of time. Assuming it is clear in the mirror, be sure that there is a sufficiently long gap ahead to go for, wait for the opportunity, move out *before* accelerating (only when you have a good view) to see how the overtaking manoeuvre 'fits' without any commitment. Give a warning, if necessary, and *see evidence of reaction.* Then accelerate, firmly if required, and lift off the power to 'run' alongside the gap, 'tuck in' effortlessly without braking and without fuss. Give a good example to others.

Here's a classic example of acceleration sense. Imagine a half mile straight of clear road leading to a sharp bend. Many drivers may lightly accelerate over at least two thirds of the straight then brake for the bend. Instead, use full power when the view is best at the beginning of the straight, over say one third of the distance. During this time assess the speed for the bend and ease off the power at precisely the right moment, so that engine braking brings the speed to exactly that required for the corner – it requires good judgement which is not possible if the speed limit gets in the way of this technique, but well worth practising if it does not. A surprising fact is that fuel consumption is the same for both styles of driving, if speed and distance is equal. This kind of skilful acceleration sense greatly assists concentration, alertness and safety.

Acceleration sense requires the simultaneous judgement of speed, hazard potential, car ability and road conditions, and is exceptionally difficult to acquire. Its perfection will demand constant practice and total application, all the time you drive. There is always more to learn; cars are different and the more powerful they are the more demanding, yet satisfying, they are to drive. It takes many years of diligent application and responsibility to match your acceleration sense to the potential of the world's great marques like Porsche. Remember that you will need patience to develop your forward planning and judgement of speed and distance to cope with great acceleration and avoid unnecessary braking. It is hard work but well worth it.

AUTOMATIC TRANSMISSION

Gearbox sense is still required when driving an automatic transmission. Although easier to handle at moderate speeds than a clutch and gearbox, it requires understanding and practical training to drive quickly. In normal driving the advantages are apparent, as it avoids the not so expert being caught in the wrong gear once 'Drive' has been selected. Leave the gear selector in this position for the length of the journey with the advantage of two pedal control and automatic variable gear selection, and no adverse effect for normal speed and level road. There is less fatigue when driving over long periods as no gear lever movements are required with the left hand, leaving both hands to steer. The power unit cannot be over-revved in the lower gears and automatic gearboxes can be more durable, because gearbox lubrication is provided from ports in the gearbox mainshaft which is hollow and oil is fed from the torque converter under pressure. The most abused component in the average driver's hands, the conventional clutch, is replaced by a torque converter. The converter seals can only be damaged by serious misuse and gearbox lubricant is changed every 24,000 miles, as against 12,000 on most manual gearboxes – a considerable advantage for high mileage in heavy traffic.

Best results can however only be obtained by drivers fully conversant with the construction of the torque converter and method of power transfer through its fluid movement. It will give twice available torque output when you press the accelerator in the first instance – real 'bite' when you take off. There are considerable practical advantages in this effortlessly smooth power transfer which comes through the torque multiplication of the fluid drive. It needs very skilful acceleration sense to 'smooth out' the build up of speed from the start to when the torque converter is engaged. Don't press the accelerator harshly, feed the power in smoothly.

In busy traffic this effortless efficiency more than offsets any slight power loss at high engine speeds. For example, consider two drivers in the same cars, apart from the transmission. The 'manual' driver would require a high degree of skill and a great deal of effort to stay with an 'automatic' driver whose foot would merely rest on the floor under 'kick down', particularly as many automatic gearboxes have four or more speeds in the transmission. The 'manual' driver would also have to cut off speed for the first bend earlier than the 'automatic' driver, if he or she is to match engine speed to make a 'clean' downshift, unless they can 'heel and toe' skilfully. It is when driving swiftly at speed like this, using the full performance of the motorcar, that as much skilful use of acceleration is required to drive fast and safely as with manual transmission, particularly on slippery surfaces.

Although for most situations you should stay in 'Drive' (automatic transmission should be used as it is designed, automatically) you gain 'partial kick down' into a lower gear for light to moderate acceleration, and 'full kick down' for firm acceleration. Don't back off the throttle and slam the foot down, rather, progressively open the throttle then flick the foot over the last half inch, to operate the full kick down. It is important to know the maximum speed of the intermediate gears, so that a decision can be made if one is beyond the range of 'kick down' for the gear that is required. Depending upon the adjustment of the gearbox, kick down should be available to within approximately 10 mph of the maximum speed of that gear. Guard against kick down when cornering in the wet and ensure that the car is quite straight and true. Skidding and/or wheelspin can occur if kick down occurs on a slippery road surface.

Those who feel insecure and say that there is not the same car control with an 'automatic' on slippery surfaces, should pop down to the Hendon Police Driving School (where John was a civilian instructor), and watch the excellent demonstration of car control by the instructors on the skid pan. They place the transmission in 'Drive' (or reverse) and leave it there for all surfaces and situations, using 'kickdown' early but when straight, to obtain speeds which are difficult to follow.

In corners where vision is restricted, as on most bends, the kick-down must be *after* the length of sight line or vision opens up and when the steering path is expanding, with a margin of safety at the exit of the bend. One should therefore select 'Drive' for all forward movement except when car control can be improved by the manual selection of an intermediate gear in three separate and distinct situations or conditions:

1. Control of road speed through and between a series of bends, provided that the anticipated speeds are within the range of road speed of the intermediate gear selected.
2. To help the braking system to hold the car back down a long steep hill. Brake first of all to the safe descent speed which is invariably *very slow* at the top of the hill, and then select second, or first intermediate gear, *not dangerously halfway down the hill.*
3. You may choose to select an intermediate gear to maintain position in a stream of traffic, thus preventing excessive use of the brakes.

When selecting intermediate gears you can, safely, lightly press the accelerator to have the engine just pulling, which will assist the brake bands in the gearbox to change gear. However it does not help to make them do so against the compression of the engine when your foot is off the throttle. Therefore, brake first and change gear separately as you should when driving gently with a manual transmission. Don't brake with the left foot – what will you do if an emergency occurs? Can you say with all honesty that your left foot is going to have the same skill as your conditioned right foot, to cadence brake for example. I suggest you place your left foot away on the 'dead' pedal and forget it. The only exception is for slow speed manoeuvring, after some considerable practice at slow speed in confined spaces.

Do not manually change gear as you would with a normal transmission, with all that wear and tear to the gear change brake bands, and if you inadvertently press the button, you may slip into neutral and back to 'Drive' or even into reverse, causing damage to the transmission and with potentially disastrous results – only change if you have to. The engine can only be started in 'Park' or 'Neutral'. This is a safety device which will prevent the engine firing with the foot partially down on the accelerator resulting in the car moving forward or backward in a confined space or in noisy traffic. *Always press the footbrake* when you select Drive or Reverse to prevent any 'creep' – dangerous if someone is walking past the front or rear of the vehicle. Therefore, whenever an automatic car is stationary for four seconds or more, apply the handbrake. It is an offence in this country to leave the driving seat with the engine running – many forget this requirement.

In conclusion, although we would choose to have a car with a close-ratio gearbox for the fine degree of car control it allows, we are the first to appreciate the virtues of the torque converter and automatic transmission, particularly in a busy traffic environment. On motorways, the type of transmission does not matter in the slightest.

STEERING

Many drivers seem to disagree with the police method of steering control, although all police training and the High Performance Course is based upon using this flowing method of altering the car's direction. Their disagreement is partly lack of understanding, partly a matter of ignoring the importance of being able to react in an emergency and, perhaps because it takes a great deal of practice to perfect. It can be very difficult to master completely particularly where it is needed most, at speed or driving fast on the skid road. In these circumstances years of practice pay in the vital area of safe driving – let us try to explain our reasoning.

British police driver training is recognised as the very best in the world, and their superb driving schools train policemen from many other forces. The expertise that their intensive programme of training imparts is, to say the least, impressive. Imagine the scene, a patrol car travelling very fast, siren on and approaching a crowded junction. Danger is extreme, the driver brakes hard to turn right, checking right, left, right, assessing the junction safe. More than anything it is his steering control that will impress you. Hands moving smoothly and rapidly, top to bottom, around the wheel, then the acceleration, siren off when the danger is over and he is gone in an instant, driving

Correct steering technique is to 'feed' the wheel through your hands. You avoid crossing your arms and always have control.

fast and safely. "That's advanced driving", difficult to copy in practice and in the method of execution. Their system of driving is probably second to none and if only more drivers would undergo training in their methods, the number of accidents would be reduced. If you keep to the driving plan you will be safe and restrained. Similar to the police system, it is not automatic and requires *training* to master completely – testing is no substitute.

Steering the police way means that you will *always* be able to steer more, and you will never cross your arms. If your arms are all wrapped around the wrong side of the steering wheel, you cannot turn the steering wheel if an emergency occurs. That is why you should not cross your arms when steering through bends and junctions and most instructors will fail to give this reason. Crossing arms usually means using the small top quarter of the wheel with frequent displacement and less contact, hands ending up in the wrong place to be able to react. It is far better for hands to be in the most effective position, at an 'equal' quarter to three in the middle of the turn.

Approaching a sharp turn or hairpin the correct technique is to steady the wheel straight, then take the hand corresponding to the turn – that is the left hand for a left turn or right hand for a right turn – over the top of the wheel, not in the turn, but in preparation

for it. This may go beyond 12 o'clock to the other side, turning the wheel 270 degrees, three quarters of a turn before changing grip with the other hand at 6 o'clock. Thereafter, hands sweep the wheel with 180 degree contact on either side, with the best control and security and sound grip from 6 o'clock to 12 o'clock using the full circumference with both hands. They should turn the wheel swiftly if required and smoothly but always in relation to the road speed of the car, without the wheel jerking to a stop each time there is a change of grip. It takes a lot of practice – try weaving between cones in a steering exercise on private ground, or a tight figure of eight timed on the watch.

The only exception to this technique is in a gentle curve and with a 'quick' steering motorcar, when you can safely retain your quarter to three position and displace your hands as little as possible. Particularly important is the constant contact of both hands with the steering wheel (without leaning your elbow on the driver's door), so that you always have full control. Remember that the gear lever is for changing gear; it is not a hand rest! It helps if you curl your thumbs over the central spoke pads, then for gentle sweeps and curves in a car with reasonably 'high' geared steering, you will be able to maintain the position of your hands; provided neither passes 12 o'clock in the middle of the curve.

We have already covered the importance of your seating position so do not forget that your arms have to 'work' the steering wheel. Watch how that well known road character 'the boy racer' sits with straight arms, whereas in reality most leading drivers (including the great Stirling Moss) sit with bent arms, and only straighten one arm when cornering. With straight arms most of your arm cannot work, only the wrist and shoulder. You must be able to apply sufficient pressure at 12 o'clock to hold the car in a long curve. Strength comes from the shoulders and arms need to be slightly bent to work properly. Although some steering control comes from the wrist, don't be a straight armed 'boy racer'. Nothing should interfere with steering control in bends; all approach sequences and braking should be completed.

'Linking' of the accelerator and steering is essential to balance the car in a bend, or particularly, a series of bends. This linking of steering control and the forces operating on the car under power, is possibly the most important aspect of car control, not least because of the link between steering deflection and power. The technique depends on whether the car is front or rear-wheel drive, and the weight balance between front and rear. We shall deal with this in more detail later.

When turning, 'flow' into the bend with the engine just pulling and maintaining the speed you have judged on approach, not increasing speed or accelerating until the bend 'opens'. A steady speed when cornering, assessed with good observation and forward planning on approach, will 'balance' the car on its suspension. This fine co-ordination between steering and power shifts weight distribution slightly to give more or less tyre to surface grip, front and rear, in a continuous balancing act between throttle, steering and speed. Always match vision with speed, keep the car level on its suspension to corner as the designer meant it to and blend steering and power on the exit as the view improves. This requires great skill and is not the oversteer, wheelspin, opposite lock technique of the race track. It will provide your passengers with a smooth ride and take the car through bends with maximum security and minimum roll and fuss, but still quickly. Again, the overall criteria are safety, smoothness and unobtrusive progress.

A 'Control' Philosophy

The primary objective of advanced high performance driving taught on the High Performance Course is,

"safe, swift, and unobtrusive progress".

Safe, sure and considerate advanced motorists always drive with 'something in reserve'. Their superb technique ensures that they always have something left in an emergency; they leave themselves 'room'. This means that they always have skill in reserve to deal with the unexpected; an alternative plan to cope with emergencies. An accident can be described as a build-up of a set of circumstances, only one of which needs to be out of control. When taught advanced driving you are made more aware, through the mind of an experienced co-driver, of the situation that may appear to be unexpected, but will be quite reasonable to the mentor. It is often not the first situation that causes a problem, nor the second, it is the third that catches you out – and the co-driver *will tell you* before it happens! With practice and proper training you will develop that extra 'sense' of awareness; it is not 'magic' but experience, acute observation, and anticipation.

We must emphasise that although the techniques described here will, when applied with discipline, restraint and consideration for other road users help you to achieve a high standard of car handling, that is only a part of the skill of a master driver. There is much more to learn yet. Practise these techniques until they become ingrained in your driving like the grain of a beautifully matured piece of wood. Essential as they are to the whole 'structure' of your driving they are only a part. We will now proceed to the other elements of high performance road driving, which will help you to enjoy your motoring. Always drive with safety as the primary objective and consideration for all other road users as the second. The objective of safety must be paramount at all times. Driving on a track is relatively easy; you only have traffic proceeding in the same direction with no side entrances (except the pit lane!). On the road you must always have that element of 'reserve' for an emergency.

CHAPTER 2

Planning Your Driving

Planning To Drive

The secrets of excellence in any field of endeavour are planning and persistence. This is no less true in driving and your planning and advanced thinking must start before your journey begins; planning the route, preparing the car, yourself and your passengers, before you start driving. Your choice of car says something about your character and judgement, a choice that will have an influence upon your safety and comfort at the speed and distance you intend to travel. Before you drive swiftly get to know your car well, by giving yourself some skid road training and taking it to a circuit (on a Porsche Customer Driving Day), and become familiar with the handling, road holding, steering and braking when driving it 'under pressure' – be confident driving at all speeds and in all weather conditions. Remember that such knowledge of your car and as importantly, yourself, is only achieved with the right training and practice in appropriate places. *You cannot carry out such training on the public highway.*

You should always familiarise yourself with a motorcar's technical specification and layout before driving it for the first time. Firstly, engine valve and cylinder layout, cubic capacity, bore/stroke ratio, fuel metering and ignition system, engine power and torque outputs at different speeds. Then the kerb weight of the car and its power to weight ratio. Secondly, its maximum speeds through the gears and acceleration times from 40 to 60 mph and 50 to 70 mph in second and third gears, for overtaking capability. Thirdly, its handling characteristics will be indicated by the brake and steering specification, weight distribution front to rear and the suspension layout. Is the car front, rear or four-wheel drive and if the latter, what type? Finally, the recommended tyres and inflation pressures; then you can check them.

Before any major journey you should carry out at least the following checks before you start your cockpit drill as detailed in Chapter 1.

The operating manual should give you all the information and more. Of course you may think this is a counsel of perfection? Perhaps, but if you *do* undertake these investigations, they will not take long and they may save your life. It is particularly important to ensure that brake and clutch fluids and the oil level reading on the dipstick are *all* above the minimum. Check the engine coolant level, and all the hoses for leaks. Use an accurate gauge to check tyre pressures and set them for load and speed. Inspect each tyre for cuts and bulges (on the inside too) and for legal tread, at least 1 mm. The Porsche engineers at Weissach recommend that your tyres should *never* have less than 3 mm of tread all over, in spite of the legal requirement of 1 mm. They have found that below 3 mm of tread tyre performance deteriorates rapidly. Naturally, always know where your car's jacking points are; they should be regularly checked for serviceability.

Finally, remember to make sure your Porsche tool kit is in place. Always carry a first aid kit, torch, and for continental journeys, a warning triangle is a legal requirement, as are a set of spare light bulbs. Naturally, for a continental journey your lights will require setting for driving on the right.

Fast and safe motoring needs a fast and safe car, so it is essential to ensure it is sound before you drive. Each and every occasion you take the wheel, particularly at the start of the day, you should compose yourself before the run and spend a little time checking the car 'cockpit' and the controls with a consistent routine.

Your cockpit drill should be simple, comprehensive and the same for any car you drive. With practice and constant use it will not be a 'chore' but an instinctive and careful element of your 'planning'. We detail our suggested 'cockpit check list' in Chapter 1.

Stowing luggage in a car for a long journey (or a short one) is an important factor in safety. Nothing should be free to fall and slide about, front or back. Make sure maps, torch and other items you will need on the journey are near to hand but secure. If you stow items on the floor make sure they are behind the seats and firmly stowed. The lower you can stow luggage the less it will affect the car's handling characteristics adversely.

Never trust the brakes of a strange vehicle but carry out a static brake test with the engine running by pressing the pedal, releasing the handbrake, and ensuring that the brake failure warning light goes out. Press really hard and if the pedal feels spongy, not firm, the brake fluid may be aerated. There is a fault in the master cylinder if the pedal sinks slowly towards the floor. It is often wise to do a proper 'running' brake test. First find a clear road, both to the front and rear, take the car up to 30 in third gear and use the mirror again to be sure! Place the ball of the foot squarely on the pedal, take up free movement to make rubbing contact of the linings or pads with the drums or discs. Initial pressure should be light, to transfer weight progressively to the front of the car and to sense any tendency to veer, then 'bury the pedal' to feel the 'muscles' of the system and see how the car responds to an emergency. There should not be any effect on the steering or violent pull, left or right, and no one wheel should lock before the others. The car should pull up without drama, skidding, or swerving about. If the rear brake limiter valve is not working, the back wheels will lock up first and you may skid down a camber. There would be less tailgating if drivers did a running brake test occasionally when it is safe to do so, as they would then be aware of how quickly a motorcar can pull up when the brakes are stamped on hard.

On any journey, but particularly long, continental or domestic ones, comfortable clothing is important, so try to wear natural fibre cotton or woollen clothing, that lets the body breathe properly. Loose clothing, no tie, and light, non-slip shoes without a welt to catch the side of the brake are a good idea. With the thick leather rim of a modern sports steering wheel, gloves are really unnecessary. For a plastic or wooden rim, wear cycle racer's gloves (the sort with the fingers missing) for sensitivity.

Planned driving gives a safe result and if the route or journey is strange to you, planning your route in your mind is essential before you start. Use the now excellent road maps to plan your route, and *never* try to read a map on the move. The main outline of the route can be jotted down as a 'strip route' on a stiff card (perhaps the night before), as you would see it on each road sign, then tape the card to a convenient place on the dashboard ready for the journey. Take note of all the dual carriageways and motorways that make overtaking easy with the exit number to turn off, so you do not miss it. Note the distance between each town and make full use of the mileage recorder with the 'trip' to record distances between larger towns and dual carriageways.

Consult your newspaper, look up on the teletext or contact the AA or RAC to find out about road works or the British Telecom road and weather services (telephone numbers in any telephone directory) and make full use of local radio to find out about local traffic conditions. Plan to stop every two or three hours for a stretch, or when you top up with fuel – guard against fatigue.

Despite varied weather and traffic conditions, with experience you will be able to estimate with great accuracy how long a journey will take you. For example, traffic currently runs through London at 12 miles to the hour. Local knowledge is your greatest asset – where the hold-ups occur and how the pace of traffic is affected at different times of the day. Know your 'patch' with all the alternative routes, but remember the shortest traffic jam is a straight line and sometimes using the back streets is not always the quickest answer. Invariably, the local council will signpost you round the town whereas the swiftest route is often straight through the middle, when it is *quiet*. Many road maps include small inset views of main towns indicating the through routes. Be flexible, investigate, and try to 'flow' along at an easy pace, with time to react to read signposts in town. One wrong turn and you have lost all the time gained. With wise use of the map you can make your journey as interesting as possible. Take to the 'back' roads occasionally, using country 'A' and 'B' class roads and motorways to avoid big cities – 'flow' along swiftly, with concentration and skill.

Across country, a 50 mph average is good, brisk driving if it is safe and should be possible if you are consistently motoring with purpose. Remember this will be reduced to 40 mph or less with an occasional short stop; you only have to pause a while for all the traffic you have overtaken to repass you. Pace your journey well for consistency, matching speed with vision, and road and traffic conditions with a highly developed road sense. If traffic is very bad do not 'leap frog' unless you have an immensely fast car and can overtake safely and unobtrusively; wait for the dual carriageway noting them on your strip route before you start your journey. On modern motorways, high average speeds are now very common place in ordinary cars, even with a 70 mph limit – such is progress.

The Driving Plan

Disciplined driving is safe driving and total familiarity with a driving plan will ensure that you have the correct foundation upon which to build your driving skill. It should limit any lack of timing or judgement in the sequence of thought or action when preparing the car for a hazardous situation. Hazards may be created by traffic, or a physical road feature which may change the motorcar's course or speed. Correct application of the 'plan' will result in the car always being in the correct position, speed and gear for safety and efficiency. It only remains for the driver to communicate with other road users with timely signals to result in a safe and simple sequence of actions for each hazard.

There are six features of the driving plan:

1. Mirror.
2. Signal before position.
3. Position.
4. Judge the appropriate speed, braking only once, if required.
5. Select appropriate gear for the speed.
6. Give warning of approach if necessary, be sure of positive evidence of reaction.

then you may safely accelerate.

The police call this sequence of action 'The system of car control', but car control is only part of the plan on approach to a hazard, which is mostly concerned with road procedure. Road procedure is all about how you select the course, speed and communicate with other road users.

The 'system' was devised by Lord Cottenham, a gentleman road racing driver commissioned by the Home Office in 1937 to select instruction staff for 'advanced' driving and advise them on their future task. He did his job very well and his influence remains to this day. One of the original six instructors at Hendon appointed by Lord Cottenham taught John Lyon at Hendon and insisted that all novices were taught the system with arm signals only, at speeds up to 50 mph. For the first three weeks of training students had to imagine that brake lights and indicators did not work!

The use of arm signals, as necessary, made a great deal of sense. It encouraged students to use their mirrors before selecting their course, before braking and then a further arm signal, if required, as confirmation before turning – you may need all of them before turning right off a busy main road in fast flowing traffic. When the 'system' was devised in 1937 'trafficators' were used for indicating a change of course. These little lighted semaphore arms were often obscured by the car's body and were inadequate when compared to the police officer's clear and decisive 'palm' signal – after all it was him giving it not some funny little contraption that was part of the car! However, things do improve and indicator lights today are usually very clear. Unfortunately drivers too often 'flick' the switch automatically, too late and as an *instruction* rather than an *indication of intent, or request.* It would do drivers no harm to go back to school and learn Lord Cottenham's 'system' with arm signals given clearly and as necessary – they would be taught to make proper use of their mirrors, give signals only if required, clearly with good timing, before any alteration of course or speed.

As a racing driver, Lord Cottenham knew that to drive fast and safely, it is essential to have discipline and he taught it well with his 'system' which was that correct positioning, speed and gear should be applied on approach to any bend or hazard. Good fast driving is only advanced thinking, applied with self-discipline and driving skill. This quality of driving stems from the attitude of mind of the driver, not just skilful handling. Forward planning starts before driving begins, the route to be taken, preparation of the car and the likely time your journey will take.

The ace racing driver thinks about his race, tests his car to perfection and plans tactics. Even when driving fast in competition, all bends are dealt with on approach with the same basic thought pattern: "What position, speed and gear do I need?" On the highway, signals must be given with the same precise timing and judgement a competition driver applies to his braking and gear changing, but when? Hazards on the highway can be identified as being physical, meteorological or human. The latter are often created by the bad behaviour of others and are rather less predictable than the former. A considerate attitude of mind towards others is the very 'heart' of good driving. Unfortunately, a result of more traffic and perhaps urban driving is that drivers' behaviour becomes more aggressive.

The master driver understands people well, hangs back, does not get involved with other drivers' bad behaviour. He or she may choose a safer route or time for the journey, avoiding the rush hour, pub closing time and weekend drivers like the plague. Being 'remote', separate from the crowd yet driving at speed so naturally and easily, it is plain to see why other drivers consider these techniques to be advanced. It *is* by comparison and it has to be taught – it is unnatural in the beginning. Graduates of our police driving schools are the finest exponents of road driving and their techniques, developed from Lord Cottenham's 'system', have been refined and perfected in the light of many years of experience. Their 'system' is the foundation of good driving and until you have mastered it, applied it with perfect judgment and timing, you will make little progress towards fast and safe motoring. This is the reason why we place so much importance on this plan in the High Performance Course.

The six features of the driving plan we use on the course are the same in principle, have the same timing and result, but have different titles – mirror, signal, position, speed, gear, warning – each applied with deliberation, perfect timing and without hesitation. The result of applying this sequence is planned safety. If the driver makes proper use of observation and mirrors, signals if necessary, arrives at a situation with the car already on course, travelling at a safe speed, having changed down and given a warning of approach if required, there is nothing further for the driver to do in the hazard but steer and accelerate away. All this preparation is done on approach, early with perfect timing. Any hesitation or delay wondering which feature comes next will be detrimental, but the logic and reason behind it will help.

The Police System

COURSE SELECTED, the first feature in the police system. It is understood that the first moment you see the hazard you *mentally* select and plan your course before using your mirror and, if necessary, giving the appropriate signal.

MIRRORS, SIGNALS AND SPEED are the second feature. At one time an arm signal to indicate slowing down was used but unless any following driver is keeping unwisely close, it is seldom necessary and may even confuse. Today, it is sufficient to give a signal in plenty of time for others to react and to see evidence of reaction before braking.

GEAR for flexible control of speed is then chosen and is the third step in the police system.

MIRRORS AND SIGNALS are feature four. An arm signal to emphasise a turning around a blind bend or on a fast road is no bad idea. So use a deliberate arm signal at this stage where emphasis is required. It is seldom needed and not to be considered as a routine, but used where necessary.

HORN, if required is feature five.

ACCELERATION is feature six. If the road surface is good, then there is no problem, but if the surface is slippery the course of the car must be straight to apply it firmly – as a routine.

On The High Performance Course the 'system' is simplified and becomes mirror, signal, position, speed, gear and warning. Now let's consider the HPC driving plan in detail.

THE HPC DRIVING PLAN:

MIRROR. Never even *think* of altering speed or position without assessing the effect it will have on other road users, and always know what is happening around you, all the time you drive. Before selecting your course, consult your mirror to assess all the action around and particularly behind you. Use of the mirror is your very first precaution that enables you to assess the overall situation.

SIGNAL. Giving yourself *reason* to signal will teach you to observe the mirror first and with all round awareness. Be sure you do not miss another road user and be aware of the position and speed of everyone, cyclists and pedestrians included, so that anyone who will be helped or informed is given a well timed signal. Use all round observation with great care, and never signal automatically. Early signalling is a statement of intention; asking others if your change of course can be started. However, most drivers signal as an instruction and just pull out with a sequence of steering, signal and *then* mirror, saying in effect to the following driver "I have decided what to do and I am telling you. I don't care if I cause you to brake!" This attitude causes many accidents on motorways and elsewhere, is self-righteous and plain bloody minded, and is today sadly more evident on British roads. A minimum reasonable time to signal before alteration of course or speed is four seconds, lengthened when speed differences are higher, say on motorways, to six or more seconds before selection of course. It is only after evidence of reaction is seen from others that the steering path is chosen, and, as early as it is reasonable to line up, deflection from the existing line applied slowly – then what about a 'thank you' acknowledgement hand wave afterwards? In the first instance, if there is no reason to signal at the normal time in the plan, keep checking all the way to the hazard with all round awareness and signal the moment you see the road user who will benefit. The exception to early signalling will be if there is an earlier turn than

(Opposite above and below) Before starting any manoeuvre know precisely what is going on around you and indicate your intention to all other road users concerned. Then, wait for evidence of reaction before execution.

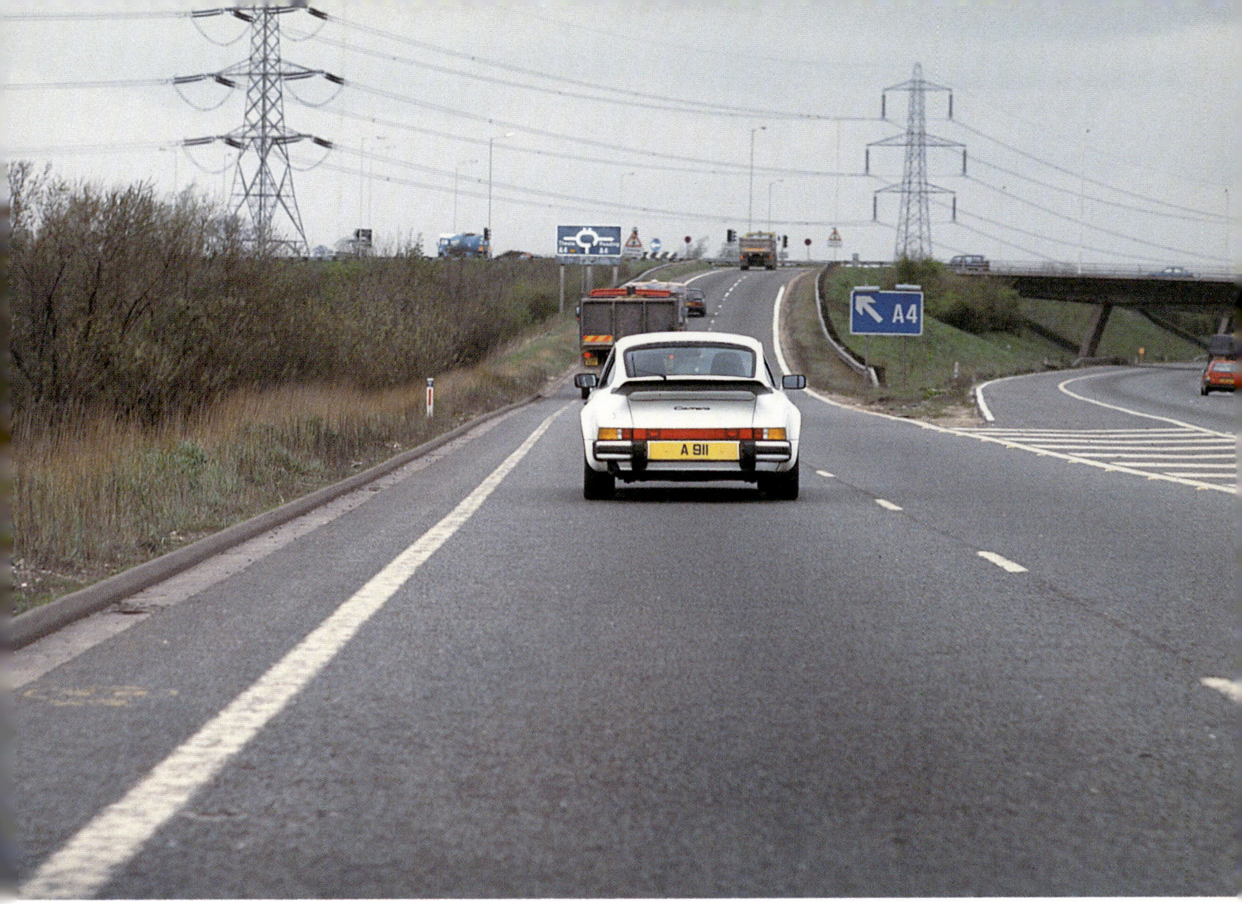
A4
A 911

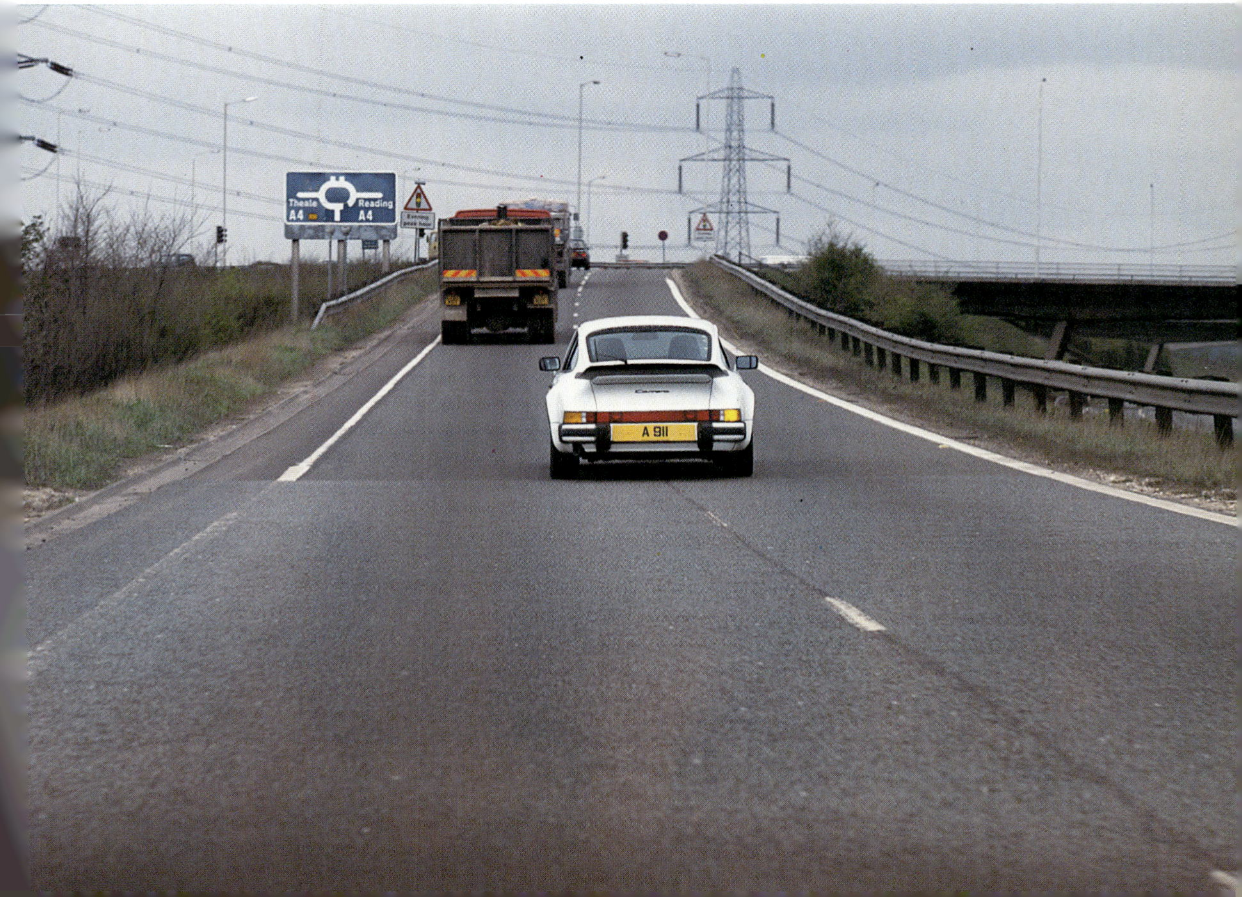
Theale
A4
Reading
A4
A 911

the one you require – it may be necessary to delay the signal to avoid confusing following or emerging traffic. Always manually cancel it at the precise time after use – don't rely on the trip.

With proper anticipation you will be able to signal early, before you cut off speed, and at least four seconds before slowing down and following the mirrors and brakes procedure of the police system. Again, look in the mirror at the 'mirrors and signals' place of the system if the situation changes during the approach. Signal sensibly, with good timing and perfect clarity. Timing and discrimination is the key to indicating – it is so easy to flick the switch on without thought. Look first, then signal if required before you change course or speed.

POSITION. Your car should always be precisely where you want it to be – positioning, like observation, must be a constant discipline. When turning you should position your car as accurately as possible to leave room for others not turning, to proceed undisturbed. Always position your car correctly, after signalling and awaiting for reaction, then take up your position. If you are early with observation, then your signal will be 'early', before you position. Early positioning gives others more time to react.

There is no better signal of your intention than road positioning to an experienced observer; but not all observers are experienced. Never forget that it is generally an offence to pass anyone on the left in this country, unless their positioning to turn is confirmed by a signal. Occasionally, timid or elderly slow drivers will travel long distances very slowly in the overtaking lane of a dual carriageway without signalling their intention to turn right. You cannot legally pass on the left, except in a one-way street. On all other roads you can only pass if the slow driver signals a right turn; although failure to signal will cause many frustrated drivers to commit an offence. Stay into the left yourself, hang back and wait patiently – you have no choice. It is very selfish driving, but you must use your experience not to frighten them, or bully with the lights or horn. This is where you can show restraint, care and courtesy to the not so able. Use your experience to help others.

SPEED. Speed must always be judged correctly and is a most important feature of the driving plan. This is where drivers display hesitation and lack of judgement, anticipation and forward planning. Many drivers are lazy, saying "I know it's a sharp bend, but I'll deal with it when I get there", giving no thought to the safe speed for the bend and the gear required. It is essential to think of the driving plan as soon as the hazard appears in view. Long range assessment of speed will enable you to 'cut off' speed at precisely the right time – to know *just* where to do

(Opposite above and below) In all situations place your car exactly where you want it; control its position at all times.

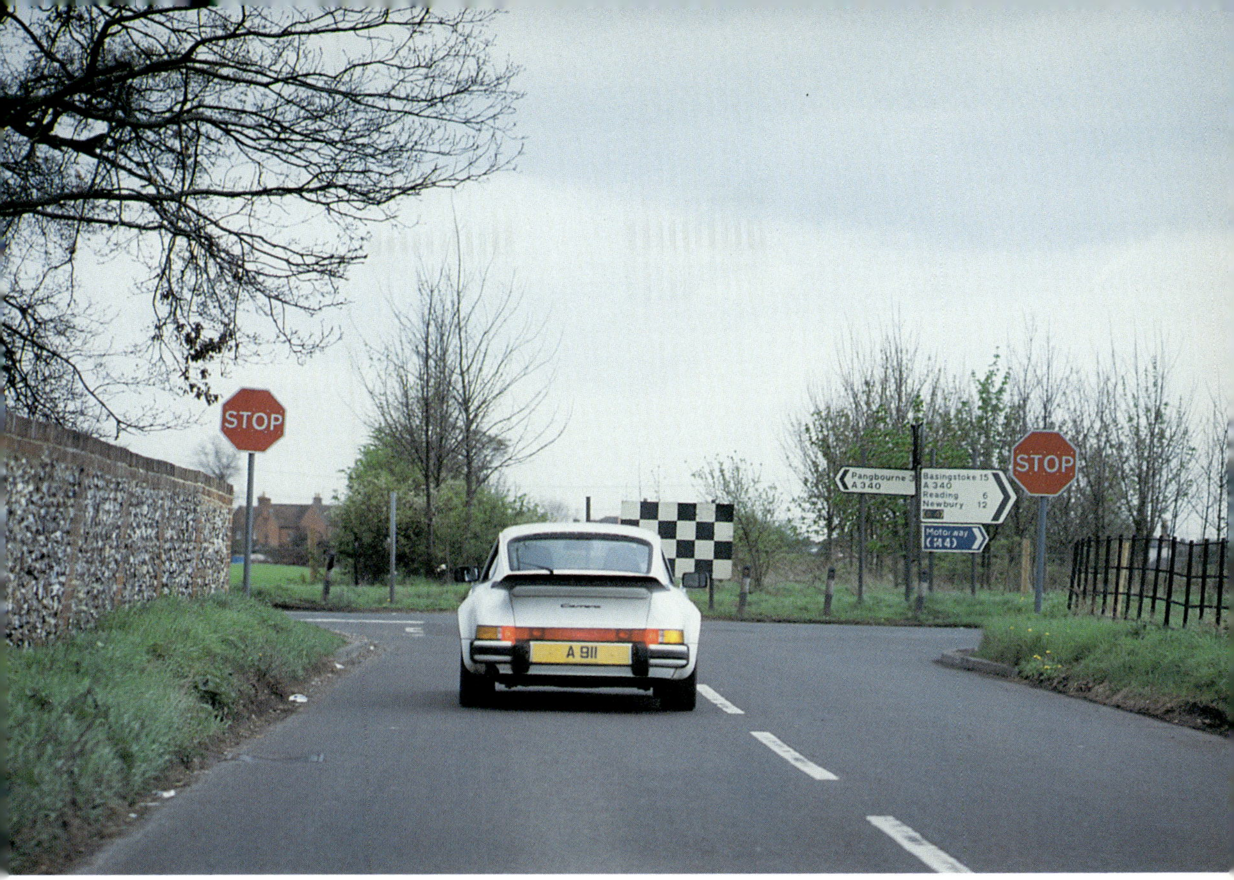
STOP
Pangbourne 3
A 340
Basingstoke 15
A 340
Reading 6
Newbury 12
Motorway
(M4)
STOP
A 911

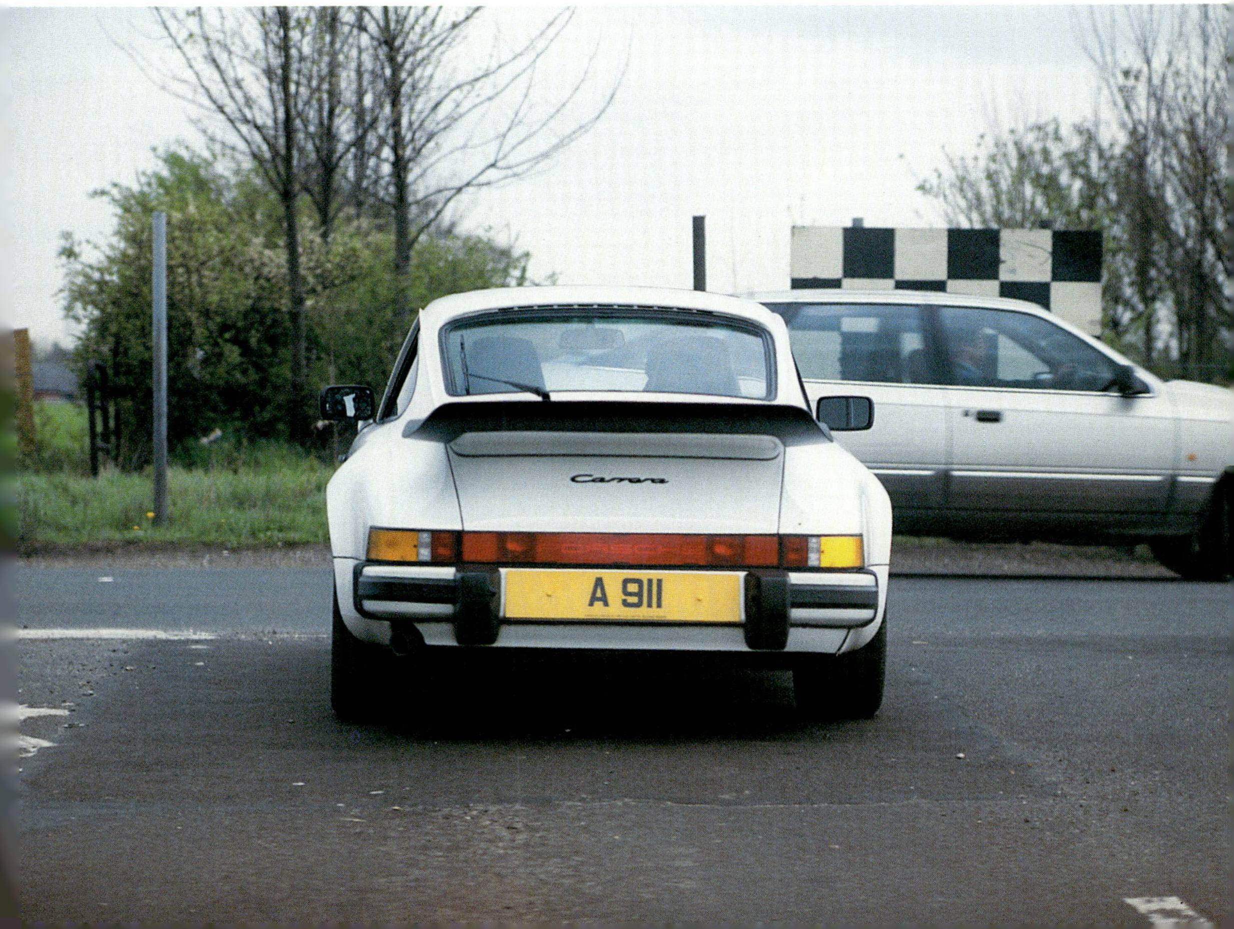
Carrera
A 911

so requires good judgement, to arrive neither too fast nor too slow. Assessment is based on the view of other road users and hazards you are likely to encounter. Remember *ALWAYS* be able to pull up and stop comfortably, well within the distance you can see to be clear – into the junction, round the bend or over the brow. How wide is the road? What field of view? Is there a camber? Is the road going uphill or down and what is the grip of your car on that surface? What degree of turn will I have to make for the hazard knowing that if I have to pull up, the car will be unstable in the curve? The eyes and mind of the good driver are busy all the time.

Judging the speed of approaching traffic when emerging is achieved by assessing the distance covered in a certain time. Distance can be judged with a quick glance, but speed varies and requires either long observation or a second glance after a time interval. You must give yourself time to look right, left, right again, as failure to judge the correct approach speed causes many accidents. *Looking* is an essential part of speed assessment in the driving plan and should not be considered separately, as in the Department of Transport's book "Driving". Why bother looking right, left and right again prior to emerging from a turning left into a major road, when many drivers only glance right then go? Well, if it is clear to the right, it's also clear to your left for someone to overtake on the major road approaching your junction; fatal if you emerge without looking left. Don't only look for cars, but for approaching cyclists, motorcyclists and pedestrians.

Many accidents are caused by drivers who look, but do not see. The final assessment of speed, before emerging, must be from a position from which you can *see*. It is useless to assess situations unless you can see what you are looking for. Yet many do. Those that do not 'see' cannot judge speed or distance.

Remember – *Look to be safe, then gear, and go.*

GEAR. This is the next feature in the driving plan. It is the control of the road speed of the car through the situation and the degree of acceleration out of it that demands which gear you need for the hazard. Until these factors are known, keep braking or reducing speed safely, with both hands on the wheel. Apply one steady brake application and change down when the road speed is safe and reasonable. Make a 'clean' change, matching engine speed with gear and road speed; double-declutching for car sympathy if required. The assessment of speed must be correct before you change down, the discipline of making sure your speed assessment is accurate by braking *once*, before taking *one* gear for the hazard will ensure that speed is *safe* and gear is correct. Do not select two gears for one situation, be deliberate and decisive.

If the distance to the hazard permits and the situation has no perceived reason to change, then reduce speed by power reduction. Link anticipated vision with speed and grip for a bend or crest and change down early into the gear you require without any braking. Some police instructors may disagree on this point, but one of the better qualities of a good driver is the ability to use the compression of the engine (via the gearbox not the clutch) to gradually reduce the speed of the car, but only if one gear change is made to gradually reduce the road speed, over the considerable road distance that it will take to be smoothly effective using this method. Judgement of speed and distance must be correct; if you brake after changing down, you have misjudged speed and/or distance.

After changing down, make another mirror check. This is essential if there has been no reason to signal before now with a clear, empty road front and rear. The scene in the mirror or ahead may now be different – although any movement in the mirror should be detected, if your driving position and mirror adjustment is precise. Look to be sure, and flick the indicator instantly if someone appears – it only takes a split second. This feature is an extension of feature two and we think it is necessary to consider it constantly – it is an on-going consideration throughout the approach. So far, the mirror has been consulted three times in the driving plan and the car is now travelling in the correct position, at a safe speed and in the correct gear for the speed.

WARNING. The last feature of the plan is to consider sounding the horn to warn of your approach. It is only to be used if it is seen to be necessary and when every other precaution has been taken. Do not sound the horn, then brake. Slow down first, change gear and then sound the horn. Use one note, either a tap or medium warning – only use a long continuous warning at high speed. It is essential to give it with time for the person you are cautioning to have time to react, and for you to see evidence of reaction before you accelerate through the hazard. Only sound the horn as information, saying "I am here", not "Get out of the way". It is the voice of the car. Use it as you would speak to someone, not shout at them. Use your left hand if the car has a centre button and you are right handed – keep your strongest hand steering the car. Curl the fingers of the right hand over the spoke to help steady the steering over bumps or cambers. Keep contact with the steering wheel rim if the horn is on the end of the stalk and 'tap'.

Don't forget the acknowledgement – a friendly wave afterwards to thank other road users for their cooperation. It does more to further good fellowship than any other signal, and is so badly needed on our roads today.

Summary: The Driving Plan

The driving plan is the discipline base from which all of your driving should grow. The features should appear to be automatic, and their application a continuous feature of your driving. The plan's elements are, to remind you, as follows:

1. Mirror.
2. Signal.
3. Position.
4. Speed.
5. Gear.
6. Warning.

These features must be coupled with the attitudes of constant observation, concentration and anticipation which will give you time to act and others time to react to you. The approach to any hazard or situation should always be a slow approach, to give yourself time to assess and react, then to accelerate out fast and safely, as soon as it is safe to do so. This slow approach and constant application of the driving plan, with a margin of safety, does not influence your overall journey time. What does influence it is effortless and safe acceleration, to reach your cruising speed quickly once more.

This pattern of driving is not natural behaviour. Most seem to think rushing hazardously into situations is necessary; it is not and is dangerous. The difference between the HPC driving approach and the norm is very marked to the observer in the passenger seat. The latter brake first and late, signal and move without giving anyone time to react, make clutch dragging gear changes. Worse still, speed assessment is wrong, then, as if to prove it, another hurried gear change is made and the hazard rushed through without time to 'see'. Accidents occasionally do happen, but they are usually created by this hazardous, aggressive and inconsiderate driving style.

Use the driving plan, develop early assessment and always act positively and deliberately. Above all be safe. Anticipate then plan your course and speed. Use signals to inform and wait for positive reaction before accelerating.

CHAPTER 3

THE OVERTAKING TRIANGLE

Unobtrusive progress is the hallmark of the master driver, the bad driver is the one we all notice. Progress that is unobtrusive and beyond the reproach of any other road user is very difficult to achieve as you not only have to drive with great expertise, skill and consideration, but you have to know other people well. That knowledge will help you to 'blend and fit' your driving into the prevailing traffic, and yet still maintain reasonable, safe progress. This requires car handling skills of a very high order applied with flair, self-discipline and above all, knowledge, anticipation and consideration for other road users. Anticipation so that you always expect them to do what they actually do, and consideration to blend and 'fit' your course, position and speed around them without confrontation.

There are those who believe that because of overall speed limits there is no need or reason for high performance cars. In our view, a high performance car is actually a safer vehicle at *all times and in all conditions* than a car with less sophisticated braking, power and suspension systems. Consider for a moment our initial proposition; "a car is no problem parked in a garage it is only when ..."

The high performance car has far greater ability to cope with any contingency than ordinary road cars. It is true that they can be used with greater danger to others by wrong attitudes, poor skill or lack of consideration. If the driver who gets behind the wheel is so inclined as to drive badly, without consideration and even recklessly the accident will, perhaps, occur at a higher speed. In the 'right' hands however, such a car is not only safer but can be driven with less fuss and more unobtrusively than an ordinary motorcar. A well designed and engineered high performance car always has that much more in reserve, can make driving easier and overtaking so much safer! For example, the crucial acceleration figures of 40 to 60 mph and 50 to 70 mph are accomplished by a 3.2-litre 911 Carrera, in 4.3 and less than 3.5 seconds respectively. Such acceleration, used with skill and consideration is safer and more convenient.

Safe, unobtrusive and progressive motoring as we have been describing is not about the excessive use of high speeds or inconsiderate and aggressive overtaking. We believe that if you possess a high performance road car you should show a good example to others and drive beyond any reasonable reproach. Remember, not many people have the opportunity to experience the sheer pleasure of driving such cars; so be considerate and safe at all times. To make safe, unobtrusive progress on today's often crowded roads requires a great deal of skill, experience, great care and consideration, as well as flair. A much maligned example is the skilled London taxi driver who will 'flow' through the thickest traffic, but knows precisely when to say "Sorry we will have to sit this one out". He is a prime example of how to use excellent knowledge (and route planning) to avoid being 'snarled up'. It takes a great deal of experience to make progress in heavy traffic, without being noticed.

Good driving starts before you sit in the driver's seat. Planning will help the 'pursuit' of safe progressive motoring, and making certain your route is appropriate for you and your passengers' needs is important. THE FIRST FACTOR is a well planned route. Then you will avoid unnecessary overtaking and annoying others by passing them just before a village or town, or when you need to take a turn off shortly afterwards, then forcing them to slow; hardly unobtrusive! THE SECOND FACTOR is to consider the point of overtaking; will it shorten your journey time or simply annoy the driver in front? Remember you are aiming to make safe, fast and unobtrusive progress. THE THIRD FACTOR to consider even before you commence the overtaking procedure is are you capable of overtaking safely and surely or is the driver in front travelling at the maximum appropriate speed for the conditions? Can your car achieve an adequate speed differential for a safe, fast overtaking manoeuvre? Never commence the overtaking procedure because you are 'in the mood' to drive swiftly, only because your answers to the previous questions are positive; then, and only then, commence the overtaking procedure.

Overtaking and turning right are the most potentially dangerous manoeuvres to undertake. Consider how often you see drivers sitting, TAILGATING, immediately behind another vehicle waiting to swing out and overtake. Think for a minute, would you turn sharp right without allowing the car in front to move ahead and open up your field of vision? Overtaking, is in effect, an uncompleted right turn and yet many drivers start that manoeuvre from a position that severely restricts their forward vision, stopping others who can see and have the power from overtaking; adding to potential hazards. When you find yourself behind such drivers beware. They dart out, signalling as they move, often into the path of an oncoming vehicle. Before beginning to overtake you should move out and await positive signs of reaction. Do not risk him, or her, moving directly into your path, because they will probably not check their mirrors before moving. Safe and responsible overtaking begins with discipline and anticipation. Remember, always give yourself room to work, move and see.

Drivers who sit on your tail, incessantly 'pushing' forward, although there is no opportunity to overtake and make further progress are a menace. They preserve the integrity of their front bumper and grill only by stamping on the brakes. Such behaviour is unfortunately often 'lemming like' and responsible for many of the problems of driving in heavy traffic. Patience in such circumstances is 'the' virtue. Undoubtedly their reasoning is the same as the 'motorway tailgater'; intimidation and aggression will make the slower driver move, or prevent another driver filling the space! Do not get 'involved', but give yourself space in front if they are attempting to intimidate you.

Perhaps worse than the motorway or town traffic tailgater is the driver who never prepares for overtaking, particularly on single carriageways. They follow far too closely, change down and 'swoop' out with power applied 'from behind', steering and signalling at the same time, giving no time for others to react. If the separation gap in front is already dangerously close (and it usually is), then 'power' from behind will shut the gap – a potential disaster if the driver in front has to brake. This overtaker only thinks of oncoming traffic and 'swoops' around the car in front, instantly committed to pass without a clear view or any warning to other road users. This type of manoeuvre is always potentially dangerous, and since it is used by the majority of drivers it is not surprising that it results in many accidents. As you have so often discovered, the natural way is not always the correct or the safest way.

The standard defence of the use of this 'power from behind and swoop' overtaking procedure, is to say "but I am limiting my time on the 'wrong' side of the road". These drivers are oblivious to the fact that accelerating first and beginning to travel past the car in front, with very limited forward vision, is moving into danger with a limited escape route; they will have to brake to slot back in behind, if there is oncoming traffic. Perhaps it is driving small cars with limited power which has dictated this "charge from behind" approach. Firstly, to cut down the time on the offside half of the road and secondly, because they think it means they have less far to travel to overtake. Both thoughts are factually correct, but totally wrong in terms of risk! If the danger is concealed when you commit yourself to overtake, the hazard will be exposed too late to do anything about it. (See diagram 3/1).

3/1
Tailgating limits vision and maximises risk.

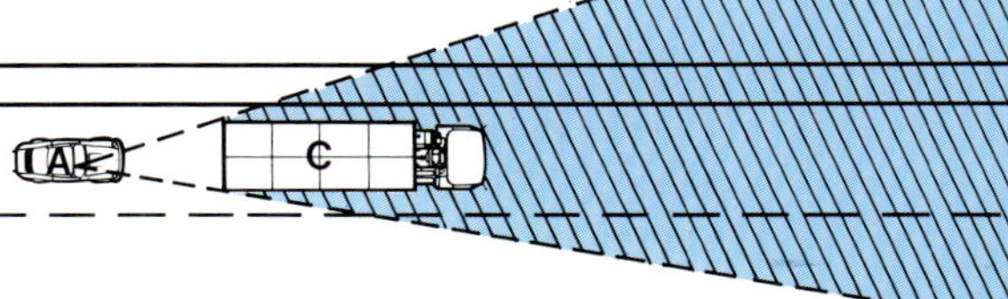

The very best view range that this technique will ever give you is 90 feet, and it will often be less. As you, the driver of vehicle 'A', swoop out under acceleration you are not accelerating in a straight line, have a greater risk of skidding and you limit the amount of acceleration you can safely use. If vehicle 'C', the lorry, is travelling at 40 mph, and you in vehicle 'A' and an oncoming vehicle are travelling at 60 mph, you will have less then 0.30 seconds to take avoiding action; your closing speed is 120 mph, and the closing distance is approximately 45 feet! Increase the speeds and the times get shorter, potential danger gets greater. Naturally, the preceding picture takes no account of the fact that from the position described you, the driver of vehicle 'A', cannot possibly see other potential hazards such as concealed entrances, vehicles or cyclists on the same carriageway, but ahead of the lorry, and so on. Such overtaking procedure maximises potential danger in every way. Let us look at the alternative.

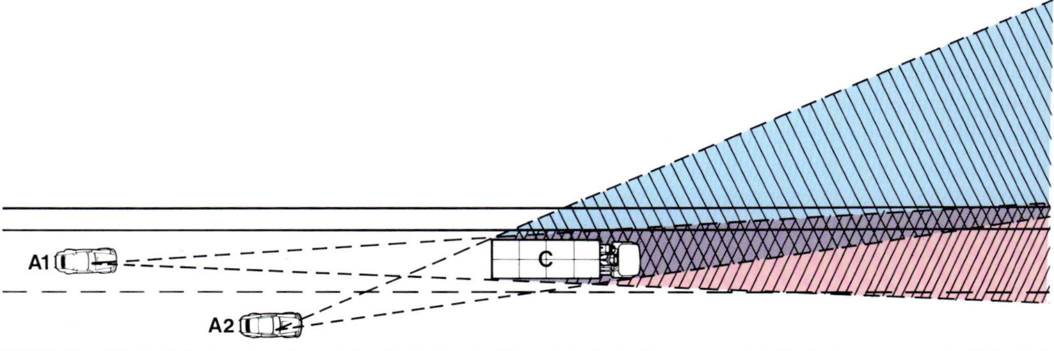

3/2
The overtaking positions: A1 the 'hang back' position maximises vision and safety, A2, the 'attacking position maximises safety and allows you to accelerate in a straight line.

In diagram (3/2), your position in vehicle A is approximately 25 yards behind vehicle 'C', in position 'A1' on the nearside. You now have a clear line of sight of at least 85 yards, some *three times* as safe in terms of sight line for the sacrifice of 20 yards. Even before you move to the offside you can now see well ahead and down the offside verge to check for entrances, pedestrians or any other potential hazards. Such extended vision could save your life.

You can then, and only then, judge correctly whether your overtaking is necessary and safe. On single carriageways overtaking itself is easy. It is this detailed, deliberate, and safety first preparation and planning that is difficult. It starts as soon as the vehicle ahead is sighted. Overtake from a 'tailgating position' and you will certainly 'miss' safe overtaking opportunities and you will unnecessarily expose yourself and your passengers to hazards and potential danger.

The overtaking procedure can be broken down into three separate phases; preparation, planning and execution. Each phase will of course be linked to elements of the six-point driving plan described and analysed in Chapter 2.

PREPARATION starts as soon as you see a vehicle ahead. Of the many questions to ask yourself are, *Question 1*: Is this particular overtaking going to be effective and will it shorten my journey time? Nothing may be gained if approaching a town or a dual carriageway. *Question 2*: Is the driver ahead likely to cooperate and does he or she know you are approaching? Are the wheels of the vehicle ahead observing lane discipline and clearly on a deliberate course? *Question 3*: Are there any road sign impediments to overtaking? *Question 4*: Is there apparently comfortable room for everyone likely to be involved, particularly ahead when you anticipate you will have overtaken.

PLANNING starts as you *approach* the point 'A1' in diagram (3/2). Do not 'charge' from too far back and commit yourself too early, instead time your approach with constant long-range anticipation and judgement of speed and distance. At this stage, if the situation regarding your preparation has changed, do not close up on the vehicle in front but follow at a safe distance for your speed. This safe following distance varies with speed and road conditions but on a dry road a yard per mile per hour is usually correct. Leaving a very short gap, to stop others from passing is driving without consideration; and may lead to prosecution under the proposed legislation for "very bad driving". If you have judged your preparation and approach properly, you will be able to take the first, safe opportunity to pass. Until then a safe following distance will provide the gap for someone who wishes to make more progress.

There are three reasons for using the 'hang back, A1' position as in diagram (3/2). Safety, view and stability.

SAFETY. You are safer in the event of the vehicles(s) in front stopping or deviating without warning. Everyone has more time to react.

VIEW. Your field of vision is increased for safe planning. The larger the vehicle in front, the further back you need to be for optimum view. Looking underneath over crests, down hill over the roof, or through the rear and front screens of cars, will give you an advantage. You can use the 'nearside' view around left-hand bends and the 'offside' view around right-hand bends.You are obviously more visible to oncoming vehicles and therefore able to drive less obtrusively, since they will not be confronted with a vehicle darting out from behind another.

STABILITY. The normal overtaking position, in the centre of the offside, will enable you to accelerate in a straight line, with the best degree of stability under acceleration. Having decided to overtake, move up to position A1 in diagram (3/2), position yourself to maximise your view using all of your half of the road up to the centre line, if safe, so that you can see into the far distance. Check the mirror behind constantly so that you never spend a moment not knowing what the situation is behind you. Vary your position, gently, to ensure that you know what is happening in front of the vehicle ahead; don't weave about. If you are back enough you will not need to hug the kerb for an inside view on left-hand bends. See the complete situation so that you can assess *all* hazards along the path of your overtaking manoeuvre, including entrances and junctions, pedestrians and traffic in front; oncoming traffic is only *one* of the hazards. This conversion of view will take a few seconds depending upon complexity, great care is required. Under no circumstances should power be applied until a complete view is obtained.

EXECUTION. Then when the view ahead and behind confirms a safe opportunity with no faster car coming into the mirror view behind, *roadsides* clear of potential hazards and you are certain overtaking will be completed before any junctions are reached, consider the necessity for a signal. If there are following vehicles, particularly any 'pushy' drivers, or fast motorcyclists, give your intention to overtake at least four seconds before pulling out and closing up to your overtaking position. Give time for evidence of positive reaction, and be certain that your signal has been seen. Not "I am pulling out" but stating your intention as the Highway Code intends. Begin this part of the overtaking procedure, at the earliest opportunity, as the hazard passes or the bend begins to unfold and the view opens out. Then you can always begin your overtaking at the beginning of the straight or as soon as the view "opens out" showing no dead ground or potential hazards, not later when your view is less and the time for safe overtaking is reducing as the speed of the vehicle in front takes it nearer the next hazard. You now close up to perhaps three car lengths, checking your mirror all the time. Do not apply power except to close this gap gradually. Apply power *gently*.

Before moving out to the overtaking position, change down to have the engine speed between maximum torque and power. The overtaking, or "attacking" position (see position A2 in diagram (3/2)), will vary with the vision required, and should be chosen to enable you to overtake in a straight line under acceleration; for maximum stability. If correct, it maximises your offside rear view as well as making you visible to the vehicle in front. You must choose your gear so that you do not have to change up half-way through. If you have to change alongside the vehicle being overtaken, you will have only one hand on the steering wheel at a time when potential danger is extreme. A gear change whilst passing also puts you into neutral, and increases the time involved in overtaking.

If the situation changes before overtaking, simply fall back and slide (don't jerk the steering wheel) over and back into the nearside to your 'hang back' position. *With correct positioning, timing and acute observation this should rarely be necessary*. This manoeuvre is achieved gently and gradually with acceleration sense rather than braking. As you were not, at the moment of aborting the overtaking manoeuvre, accelerating or 'charging', but simply holding station like an aeroplane in formation, just back off the accelerator.

From the 'attacking' position A2 in diagram (3/2) you are on course for straight line overtaking, past vehicle C. So, when you judge it is safe you move right, without the application of power committing you to pass, into the centre of the offside of the road to A2, giving you an extended 'safety line' view of the road ahead. A good position from which to judge the length of the 'long vehicle' you are about to pass. However, this movement into the overtaking position must be preceded by acute observation to ensure that the offside is clear of potential hazards such as entrances, pedestrians likely to stray, junctions, as well as oncoming traffic.

This position increases your ability to judge the overtaking manoeuvre, particularly past a 'long vehicle', as well as maintaining your 'option' of returning to the nearside, easily, smoothly and safely without braking. Let us try to explain using an optical illusion. Take a pencil and point it end on to your eyes, then place it slightly to the left side, notice how easy it is to judge its length. You give yourself that and many more optical advantages when you take this, apparently, unnatural offside position from which to overtake. Positioning yourself in this manner described, you will effectively be tracing the base and adjacent side of a triangle. This is the method of overtaking taught at all the police driving schools. In our view that is recommendation enough but in more than 25 years of instructing this procedure John Lyon has found no reason to alter it in any way.

Imagine that you have sighted a car ahead and let us go through the procedure once again, fitting in the six-point driving plan.

1. MIRROR. As you approach the vehicle make proper use of the mirrors, having gone through at the least the following preparatory questions;
   Will overtaking be effective?
   Will the driver ahead cooperate and does he or she know you are coming?
   Is there comfortable room for everyone?
2. SIGNAL. You may need to signal to the traffic behind, particularly if there is a fast car or motorcycle following that you anticipate may want to overtake you.

This completes the preparation phase and you will now be in the position to assess the situation ahead and behind for the planning and execution phases. You have already decided overtaking is possible as your observation of the road ahead shows no oncoming traffic that will be affected, no 'dead' ground, entrances or junctions in your overtaking range, and no traffic preceding the vehicle to be overtaken that will affect your course.

3. POSITION. If you judge that the overtaking manoeuvre is safe, you will proceed to position A2, in the centre of the offside, if necessary.

4. SPEED. Judging your speed to overtake the vehicle 'C'.
5. GEAR. Before positioning out you will select the appropriate gear to overtake, depending upon your car and the speed of the vehicle to be overtaken. Remember the principle of maximum torque and power, and take the gear for the best control and acceleration, before positioning out to the offside.
6. WARNING. From your attacking position A2, you can now see whether overtaking is on; this is the safe 'go or no go' position. You can always slot back into the 'hang back' position without braking; as you will not have applied acceleration. Now, do you need to confirm your intentions to the vehicle in front, and give a warning? It may be essential, and always see evidence of reaction before you go.

If you have completed all of the preceding stages, and the situation is totally safe to your satisfaction, then apply firm and positive acceleration in a straight line.

Using this procedure you *always* overtake with maximum *vision* and *stability*, but you never commit yourself to overtaking without assessing all the potential hazards, not just oncoming traffic. Also, you always overtake in a straight line and you can apply maximum acceleration, if necessary, safely with the best stability. Under hard acceleration and on a curving path your vehicle is potentially unstable. Accelerate a powerful car 'from behind' a large vehicle on a curving path while passing on a wet, steeply cambered road, and it could well skid and spin so fast that you will wonder which way you are going!

Do not worry about any driver attempting to fill in the space on your nearside from behind, when you are in your 'attacking' position safely on the rear offside of the vehicle to be overtaken. In more than twenty five years of use we have never had an example of that occurring. Your positive road positioning without power will have convinced other drivers of your intention.

Positioning out to see first, before commitment, gives the same optical advantage as when judging gaps in a line of traffic. However, do not position yourself out unnecessarily when you do not want to overtake, or when observation from the initial 'hang back' position suggests overtaking is doubtful; be economical, steady, positive and deliberate. If your 'hang back' position is correct you will not have to position excessively to see, just a gentle alteration of course.

(Opposite) The overtaking sequence as you, the driver, will see it. From the 'hang back' position you have clear vision. From the 'attack position' you can accelerate safely and smoothly past.

Use the 'hang back' position to ensure that the offside is clear of any entrances or junctions; the latter should be indicated by signs. You will also be in a position to see concealed entrances, where a driver may consider that the potential danger is from his or her right, and swing straight into your path. Your assessment from your proper 'attacking' position and your original hang back gives maximum opportunity for you to spot entrances, and for the other driver to see

THE 928S

you. Applying power 'from behind' gives little opportunity for drivers emerging from either left or right ahead of you, to see you or assess the situation, with time to react! The offside of the road is not a 'no man's land' to be in and out of as quickly as possible, without proper observation. Application of the 'rule of the triangle' gives you time to assess the reaction of other drivers who may be affected by your actions. You have time to be sure of positive reaction to your positioning or signal. How can you expect the vital cooperation of the driver ahead if he or she does not know you are there or your intentions? Imagine two vehicles, one behind the other and both applying the charge and swoop procedure!

The manner of your signalling is important. Indicators are difficult to see in the mirror and it is far more effective and safer to sound the horn to the driver ahead; do not use the headlights if the warning is within earshot and the horn can be heard. Only use lights on a single carriageway if you are giving a warning at long distance and then use them properly with a 'four-second' light, do not 'flash' them quickly. It is best to warn with your headlights for fast progressive overtaking with a long clear view from well back when they will more effectively attract attention.

When approaching many vehicles in a queue, who show no sign of wishing to pass vehicles in front, it is more effective to attract the attention of other drivers by using your headlights with a single four-second flash on main beam. Overtake only one or two at a time and do not rush from too far back. Use your sense of acceleration, speed and distance, and supplement the headlight warning before commitment, with a tap on the horn if necessary, when within 'earshot'. As a rule the horn is less effective in dead sound conditions, high speed and rain reduce the use of an audible warning.

In executing the 'rule of the Triangle', none of your movements should be dramatic or sudden. Use smooth, deliberate and controlled steering and acceleration, light deflections of steering and gentle application of power to move you into the overtaking position with an 'easy' style. Then, if the situation is safe and clear, use firm progressive acceleration, avoiding harsh power application. You do not need to split the movement into distinct parts but you should, with careful anticipation and planning, be able to coordinate the steering deflection with acceleration/deceleration sense, to 'flow' into position. Then blend and grade your application of power for firm straight line acceleration as the view and stability of the situation is confirmed. There are occasions when you will have to select a course well out on the offside to avoid 'cutting in' on the overtaken vehicle. With firm and progressive application of the acceleration of a powerful car you can be rapidly clear of the overtaken vehicle. Make full use of the power of your car in such situations, when safe to do so, using the tachometer to avoid over-revving; select the correct gear!

When overtaking it is safest to use the middle of the offside, and don't 'shy away' from large vehicles. If you line up properly, give warning, if necessary, and wait for evidence of reaction (by the position of the lorry's road wheels relative to the centre line), and then accelerate firmly, leaving a margin of safety on both sides you will pass easily. With the wheels of the vehicle to be overtaken clearly to the left of the centre line you know you have the whole of the offside to use, safely. Lorry drivers or drivers of smaller vehicles may need to move towards the offside if there is a motorcycle or push bike, so you must *know* the situation ahead before committing yourself. You will only do that by strong positioning and accurate, early observation.

When using the 'rule of the triangle' to slip by a well spaced out queue of traffic, do not drive into a narrow gap causing the overtaken driver to brake. He or she may try to make it more difficult for you by keeping close, or worse still, closing up as you overtake; shutting the gap. Inconsiderate, dangerous and stupid driving as it is, you should make sure you do not place yourself at risk. Only use just enough power to pass one or two vehicles so that you arrive alongside the next safe, long space and slide in without braking or causing others to do so. Stay out if it is clear and you intend to overtake again, giving a precautionary warning, if necessary, and then slip by with acceleration. 'Pick' your way past one or two at a time. Remember, the first opportunity is the safest, second and third attempts are never quite as safe.

Overtaking by using the 'rule of the triangle' in the phases of preparation, planning and execution, will enable you to be both fast and safe. But it is the preparation and planning that are the most difficult, requiring acute and precise judgement of speed and distance, as well as detailed all round observation. Be guarded as there is a tendency to forget about the hazard *after* overtaking and you will arrive at it too fast. Concentration, anticipation and planning will help avoid this danger. Do not change up after overtaking, if a hazard is imminent and requires the same gear, but back off the power gently, as soon as you are well clear of the overtaken vehicle. Always assess your overtaking along with the overall road situation, the effect it will have on others, and on your journey time. To overtake then brake harshly to turn left or right is one of the worst examples of 'very bad driving'; and yet how often do you witness it!

Overtaking on dual carriageways and motorways without opposing traffic should be relatively easy, but many situations occur where the potential danger is not anticipated – accidents happen too frequently. If possible avoid overtaking vehicles in the nearside lane, when you are in the middle lane and other vehicles are also passing you in the overtaking lane, making three abreast. Particularly if the vehicles are large or the traffic lanes are narrow, as in the middle of a left hand curve, as you would then be almost 'blind' with vehicles each side; if one of them swerves there is nowhere to go!

Whilst on motorways and dual carriageways your lines of sight are long and you should not have to worry about oncoming traffic. Do not 'stand off' on the off-side rear of the vehicle you intend to pass. If the driver reacts naturally by braking and swerving right to avoid an accident, you may well be the victim. Make room for yourself and others by hanging *well* back and choose your time to pass, after giving time for the driver in front to see and react to you. Use the driving plan all the time, but with longer range anticipation. Your signals should also be appropriate for the higher average speed of you, the overtaker, as well as vehicles behind and in front. Separate and direct your signals onto specific targets, headlight warning ahead and indicators to slightly faster, (or simply following) traffic behind. Lengthen all signals as far as timing and distance to at least four and preferably six seconds. *Naturally you will always use signals as an indication of intent, not as a last minute instruction.*

Motorways and dual carriageways have space and long, long, sight lines; use both as much as possible. Hang well back and watch vehicles far ahead, particularly as you approach an entrance or exit. Remember not to *tailgate* but follow well back, not using the headlights if there is nowhere on the nearside lane for the preceding vehicle to go. Apply the first part of the driving plan very early so that your approach speed (speed minus speed) will not catch the preceding driver unaware. Always be aware of vehicles closing up on the nearside, so that you can anticipate what is going to influence the vehicles directly ahead of you. If there is a lot of joining or leaving traffic on a motorway junction, then take extra care, hang back and give yourself room; most accidents occur at junctions. Vehicles joining will force some of the preceding traffic to change lane to the right, as they fail to 'blend and fit' with the traffic flow, so don't try and 'force' through an overtaking manoeuvre. That could be 'very bad driving'.

Lane discipline on motorways is generally awful. How often have you travelled down an empty motorway in the nearside lane, mile after mile, only to be confronted by a lone vehicle obstinately in the centre or the official (offside) overtaking lane? Worse still, drive down the M4 at peak hours and witness the empty nearside lane and the packed centre and overtaking lanes! Spread out and always obey the Highway Code on lane discipline. There are really four simple principles:

1. Always drive in the centre of the lane you are using.
2. On two-lane motorways and dual carriageways, the *normal* driving position is the nearside (left-hand) lane. The right-hand (offside) lane is for overtaking only.
3. On three-lane motorways the *normal* rule still applies – keep to the left in the nearside lane. If there are so many slower moving vehicles in that lane that you would be moving in and out repeatedly, you should stay in the middle lane. Drivers of heavy

(Opposite) Blend and fit in with traffic, flowing along unobtrusively.

Position in the centre of your lane, and use the nearside lane if it is clear.

goods vehicles or those with trailers *cannot* use the right-hand, offside, 'overtaking lane'. Therefore do not stay in the middle lane if this will prevent HGV drivers from overtaking slower traffic in the nearside lane.

4. NB. The (offside) right-hand lane on both dual carriageways and motorways is for OVERTAKING, it is not the 'fast' lane. Do not stay in it any longer than is needed to, overtake and then move very gradually back in again, all with a good safety margin. 'Point' your car towards the lane you require, don't deflect the steering sharply at speed.

Safety, courtesy and long range anticipation are of the utmost importance on motorways and dual carriageways. Therefore it is important to hang back and always observe the safe following distance and avoid the need for braking, if at all possible. Any changes of speed should be gentle and deliberate. At motorway speeds remember to spread out your driving plan, using the space and sight lines to increase distance between yourself and other vehicles, safely. 'Bunching' is a hazard that too often ends in disaster; and not only in poor visibility. When changing lanes apply gentle and smooth deflection of the steering only after planned signalling and positive reaction by following and/or preceding vehicles. Never rely on assumptions on motorways (or anywhere else for that matter!), the speeds are too high for doubt.

If everyone observed these basic rules, passing other vehicles would be easy, but they don't, do they? So never let complacency or relaxation dominate your motorway driving, always be alert. If the vehicle you are passing is also closing on another vehicle without any corresponding signal, communicate early and positively, so that he or she is aware of your presence; give yourself room for others to make errors! That driver may be concentrating ahead rather than all round and may signal and move out simultaneously; don't place yourself in what could develop into a very dangerous situation, particularly if the speed differential is considerable. Slacken speed, signal early with lights for at least four seconds or until you see evidence of reaction. Then accelerate decisively.

Whether overtaking on single, dual carriageways or motorways, you should never close up so fast as to 'bore' through, or be forced to brake and retire behind the vehicle in front, or brake as the vehicle preceding you in the left-hand lane pulls out to overtake; you having failed to communicate. Your judgement and anticipation are both at fault in such situations. Remember the expert driver is always tolerant and considerate, so as to 'blend and flow' with other traffic, without drama. The expert driver will rarely need to brake in such situations because he or she will have used the driving plan and their acute anticipation, judgement and observation. This often means that

Anticipate your manoeuvres, use the driving plan with consideration for other road users.

A 911
A 911

the expert driver is able to anticipate the reasonable and even unreasonable actions of others. The secrets are anticipation, the skill of using your car's performance, not committing yourself too early or without receiving positive evidence of reaction and knowing your observation has been complete, leaving nothing to chance.

(Opposite) After seeing evidence of reaction from other road users, complete your manoeuvre smoothly and with correct positioning.

In all overtaking situations positioning out to look, on the course you require to pass, is *safe* when closing up since you are then clear of visual interruption, with best view and stability as you accelerate past. Progressive overtaking is easy with early selection of course, but potentially dangerous if you commit yourself from too far back or 'swoop' out too late without vision or stability. Changes of speed can be applied with precise timing and acceleration sense, if executed early with forward planning. Consider braking in such situations as an admission of 'error'. Then you will ensure that you give yourself 'room to work' as you time the close.

Overtaking manoeuvres should be carried out on a course through the middle of the offside lane, with a safe margin of error. Lorries on a motorway will cause a 'bow' wave of air that will push and draw a car at high speeds. As you pass a lorry, the slower bow wave of air will 'drag', and then release as you pass. Do not overreact to this but hold your course; always have both hands on the steering wheel and anticipate with *mild* steering correction.

If you sense any instability with a loss of tyre pressure or, worse, if a tyre bursts while passing, operate the driving plan. Mirror, signal and remove speed as early as you can by coming off the power completely. But keep off the brakes as instability will increase under harsh braking. Concentrate on steering until you are out of trouble with the minimum amount of evasive action; keep a healthy control of your nerves which is essential in an emergency.

Remember our repeated comments about anticipation, judgement, speed sense and never committing yourself until you have positive evidence of reaction to your signal. Use the 'rule of the triangle' and fit it to your driving plan, at all times.

CHAPTER 4

ADVANCING OBSERVATION

Co-driving, or simply being driven by expert drivers is a fascinating experience, not only because of their gentle, smooth and precise handling of the controls, but also for their mental ability to anticipate ahead. It often seems that they have an uncanny 'second sight' or 'sixth sense' that enables them to visualise difficulties *before* they arise, or are otherwise not apparent to lesser mortals; to 'sense' danger when no physical signs exist. The secret lies in their acute visual sense plus experience which, when fully developed, allows the expert to know the 'percentages', to assume danger and react before it is seen.

The foundation for this acute and life-saving sense of observation and anticipation is your eyesight. Of the five human senses of sight, hearing, touch, smell and taste, the driver's most important asset is sight. Your eyes are the vital link between your brain (the 'motor' of response) and everything around you; 'seeing' that is not simply 'looking'. Watch the eyes of a very good driver or pilot, notice how they flick around and never seem to fix on one point for more than a second or so. They observe like a hunter, minute detail is noticed and focused on if it is relevant. Assessment of the value of detail seen is instant and potential danger is recognised immediately, after so much driving experience. Perfect eyesight is essential particularly at night or in bad weather. When driving, search all round for detail at the relevant points of focus and far onto the horizon, make your eyes work – you should be able to read a number plate at one hundred yards in good daylight. Remember our comments on eyesight in Chapter 1, and if in doubt, have your eyes properly tested. Better still have them tested regularly, especially if you are over forty. Recent research has indicated that even if your eyesight is good in daylight, your night vision may be impaired; be warned.

Assuming that your eyesight is perfect, or properly corrected, what should you be looking for? This acute sense of observation, of looking and seeing, requires intense concentration and knowing what to look for, as the following examples may show you. Try to look at the people inside the cars as you approach; don't just look at their faces, look at their eyes, see them looking at you. Watch the tilt of their head, the facial expressions. Notice how they hold the wheel; are they about

to turn it and have they seen you? Are they in the correct position or have they strayed around in their lane, with their offside wheels sometimes over the line? Are parked vehicles empty or is the driver in place and exhaust gases visible (not from the driver but the exhaust!)? Watch for brake lights on parked vehicles ahead, and not just one ahead but as far as you can see. Position for vision; more about that later.

Safety is often the difference between seeing, and merely looking. To assess hazards you must see them and evaluate their importance.

Watch a pedestrian's feet when walking, the toes turn out when about to turn onto a crossing, the head turns to look around and they will lean forward a fraction before they run! A cyclist will wobble uphill or in crosswinds, and often stop pedalling just before turning round to look over his shoulder to see you, before deciding to signal to cross your path or wait. Cyclists can see potholes and gullies that you may not notice and they are influenced by your 'drift' as you drive by, so give at least as much berth as you would a normal car, more if you can. All cyclists are entitled to their wobble and the wobbling distance is considered to be six feet, i.e. room to fall off and spread across the road. After all they are balanced on two skinny little tyres! Remember in the wet that it is a traffic offence to splash pedestrians or cyclists, so watch out for puddles and standing water. Notice the ears of horses; laid right back they are upset, set forwards they have found something of interest, *you* maybe, and could shy when you come fully into focus. Horses are shortsighted and difficult to deal with, they take a dislike to the most peculiar things like prams, taxis, umbrellas, or anything unusual. Always try to keep well away and drive slowly and quietly by and then the rider will invariably thank you. In our opinion, equestrians are the most polite of all road users. If we were all to behave as well, there would be few problems on the road – the value of training at a riding school perhaps?

Livestock and horses will often be a hidden problem on country roads, so do not use harsh acceleration until you have sufficient vision. Sheep, when calm and occupied, will keep their head down, but look out if heads lift up and something takes their interest – maybe other sheep on the other side of the road. Anticipation is achieved by having a healthy and intelligent sense of the worst that may happen. It is relatively easy to assess what an animal is going to do, but people may intimidate you into believing that you must do what they wish you to do, for their benefit. Try not to be easily perturbed by intimidations of this kind; easier said than done, but with positive, deliberate driving you will not be 'forced' into danger.

It is relatively easy to anticipate the action of a road user you can see, but rather more difficult to anticipate those of a road user you cannot see, but who you quite reasonably expect to be there. 'Blind' entrances to houses or garages for example. Most people drive in lazily and reverse out, placing at least six feet of the rear of the car into a narrow country lane before you, the approaching driver, can see. A traffic offence, but let's face it, most people are 'nosey parkers' aren't they ? Therefore, expect it since it is quite reasonable for this situation to occur on a number of occasions throughout the day. So when you see a blind exit/entrance choose a course giving a sufficient margin of safety from the entrance. If you cannot give sufficient room because of oncoming traffic, then take all reasonable precautions, by slowing down, changing down for acceleration afterwards, and give a warning of approach. Essential in this situation as it is better to use the horn than to have an accident. Remember it is 'blind' entrances on the left-hand side that are the most difficult to see, and the ones you should consider sounding the horn for. Entrances on the right side of the road have the margin of safety created by the offside. But remember you *must* see the 'blind' entrance and not overtake on the approach to it because a minority of people will not look left first. They think that, because it is clear on their right, it is the only danger they have to consider, forgetting that if it is clear to their right, that side is empty for someone to overtake. If you miss seeing the blind entrance on your right and attempt to overtake, there is no escape for you alongside a lorry; just death if someone emerges and hits you. Detailed observation is linked to experienced anticipation of what it is reasonable to expect and only apparent to someone with a healthy imagination!

You should always be able to stop within the distance you can see to be clear. MATCH SPEED WITH VISION. Your ability to stop should always be matched with the length of your sight line, appreciating the

grip of your car on changing road surfaces. Concentration and continual all round awareness far ahead and behind you, are essential. Base your driving plan on what you can see, but also on what you cannot see but you should reasonably expect to occur. Change your focal points, at least every two seconds. The limit of your eyes' focus should vary according to speed. If your attention is required in the foreground or at the sides of the road for concealed areas of danger, then it is clearly dangerous to drive too fast and your speed must be kept low. *Wrong* places for speed such as down the high street, or narrow 'blind' country lanes, often necessitate very low speeds for safety, often well *below* the 30 mph limit.

In the country beware of pedestrians. Always give them plenty of room whilst remaining on your side of the road. In this instance John Lyon slowed right down because he could not give adequate clearance without crossing the warning line. Vision was inadequate.

Healthy pessimism may well save your life. If you cannot see don't optimistically assume that the road ahead is clear, one day it won't be! There is a potential danger around every 'blind' entrance, dip or crest, and bend. Confidence and optimism are achieved when you know by experience what others are doing and what to expect. When you can see far ahead to the horizon with no concealed danger then speed can be seen to be both reasonable and safe. Forward planning is essential, with the driving plan applied without hesitation and with perfect timing. Have a decisive and defensive attitude of mind and you will avoid 'moments'. Guard against fatigue, otherwise recognition and assessment of situations may become late and inaccurate. Practise driving for long periods; intense mental concentration, 'pin sharp' eyesight and physical fitness will all help. Experience will help you to sense if you are becoming inefficient. Then you should stop and rest quite frequently especially if you are unused to driving far; two hours is long enough to drive without a break if you are not used to long hours at the wheel. Concentration linked with acute observation and total application, to the complete exclusion of anything else, will improve your driving ability.

Without concentration your ability to observe detail and assess its value is reduced; and a little inefficiency in this context can be disastrous! Plan far ahead at high speeds and always vary speed against shortening or lengthening focal points, or 'sight lines'. Short sight lines need slower road speeds to give you time to deal with hazards without becoming involved. Vary speed with acceleration sense to give you time to react without unnecessary braking, so that you 'flow' along avoiding the need for sharp evasive action. Focus your eyes according to the speed you drive, linking vision with speed and grip. Dips in the road are potentially dangerous and are 'blind' until you can see all of the road surface throughout the dip. 'Dead ground' is the army expression and perhaps relevant if you ignore the simple advice not to accelerate or commit yourself to passing until you see all of the road surface.

Driving a safe distance behind others and never 'TAILGATING' will give you plenty of opportunity to see potholes, puddles and objects in the road and alter course very early on. Holes in the road do *not* move, they are *there*, and there is no excuse to hit them. So hang well back and plan ahead. Do not run over or hit any object with your tyres; a paper bag may have a brick in it and a plank of wood may have nails to give you a puncture – the faster you drive the more difficult it is to see them. This is another obvious folly of 'tailgating' as the vehicle ahead may 'jink' and leave you to hit the obstruction with no time to react. Range your eyes across the countryside and notice its general shape and configuration to assist your forward planning, 'reading' the countryside ahead to decide where your sight lines are going to be lengthened and shortened by hills and dips, curves and bends. Exceptionally sharp bends will be signposted, the more chevrons the sharper the corner; but never a sharp crest (or vertical bend) that will also shorten your sight line – there may be a herd of sheep over the brow! Wooded countryside will shorten your view, cast shadows to keep the road wet, and drop leaves to make surfaces slippery. A shadow will indicate the presence of a vehicle only shortly before it appears. Shadows will change your view dramatically so try to anticipate the sun as you change direction, particularly in winter.

The contours of the landscape will have the greatest influence upon gearbox and acceleration sense and you should anticipate and plan far ahead to keep the car 'working' and flowing along. Around the 'back' of a hill, in the shade, the road surface in winter may be frozen. Follow an unbroken line of trees for probable road direction, but remember that those 'terrible' telegraph poles will sometimes 'march across country', but the road will probably follow contour lines and go round the hill, climbing or descending gently.

Church spires and other obvious 'signs' in the distance, mark features such as the probable location of a village. Villages are usually easily seen and demand an early lift off the power to arrive at the speed limit without braking. The limit sign will often be concealed around a blind bend, to make you feel uncomfortably guilty if you fail to see the buildings in the distance! You will often see much more on the approach to a hazard, say looking down from a hill, than you will see when you arrive. *But if you don't look you will not see* and you need experience and knowledge to know what to look for. Always try to take full advantage of open spaces and breaks in the line of buildings, hedges, walls or fences to obtain that brief but invaluable view into converging roads, which to the less observant appear to be totally obscured. Vision into a minor crossroads that is blind will only open up quite late as you arrive. Potential danger from such hazards can appear from all directions and you should be able to stop before the danger zone in the middle of the junction. Here again you will need constant and consistent concentration on the application of each feature of your driving plan, and perfect timing to apply it in the appropriate place. Before applying a positive decision to alter course, accelerate, or turn, wait until your field of vision widens. This is an example of anticipation and forward planning at its best. Search the road surface for clues of potential danger, black lines of a skidding car indicate trouble in the past, and a time to be a little more cautious than usual on approach to a 'blind' hazard. Animal droppings or damp mud from a tractor are relevant, and if seen at an entrance you should slow down and position your car away from the it; right away if the offside is clear and there is no 'dead' ground. Similar signs on a brow should make you 'back off' more than usual. Expect the unexpected, it is often quite reasonable if you interpret the 'signs' correctly and quickly.

Constantly position your car to extend your view, without overdoing it, and 'lean out' away from an entrance to see more and have a margin of safety. But never cross a warning line unless it is clear as far ahead as you can see, and with no dead ground or other potential danger. When you spot another road user in the distance, try to imagine what the driver is thinking about. What motivates their behaviour, is it road safety, time or money? Is he or she linking speed with vision and driving up to the speed limit when safe and clear, or cruising at fifty, conserving fuel? If so you may wish to overtake when it is safe to do so. This driver's presence on the road ahead should influence your thinking and action as soon as their car is in your sight line, and the implementation of your driving plan should start *then*. As you reduce the distance between you, maintain close observation of the car ahead and constantly 'range' your eyes between it and left, right and behind (using your mirrors), to help your judgement of relative speed and distance. If you decide to overtake do so at the first safe opportunity as part of your driving plan, and only at the first, certain, safe opportunity. Maintain observation of the car's road position (definitely and constantly to the left of the centre line?) and speed (slowing or accelerating as some drivers do as a faster car approaches?), and most importantly has he or she *seen you?*

Implementation of the driving plan should be a constant, disciplined, and logical process which must be learnt. You will find that the more experience you gain in driving, the more you will appreciate the value of self-discipline as an aid to developing driving skill. Well developed skill and early discipline are two essential ingredients in driving expertise – ask any great racing or rally driver and we are sure they will agree. A very effective method of helping that development, is the use of the driving commentary, at all speeds, weather and traffic conditions. Driving should never be boring with more to see and understand. A fluent, articulate commentary that blends well with your driving needs lots of practice and requires total concentration, to the complete exclusion of anything else. It should encourage your observation and anticipation by forcing you to explain the action you are about to take to deal with the many hazards of the road, *before* they occur.

As you may feel self conscious at first, start by yourself and practise by identifying all traffic signs and road lines giving them their correct titles; and if you don't know them, buy the D.O.T. book 'Know Your Traffic Signs' and swot them up – discipline starts by gaining knowledge. Mention each road sign as it appears in view, don't only give its correct title, but say what it means to you and what you intend to do about it, if anything. Eventually your self consciousness will disappear with your improved knowledge of sign observation and your commentary will have confidence, fluency and will literally tell the story of your progress. As your proficiency increases, longer periods of concentration can be undertaken while you gradually increase the content. As with your use of controls, be precise and economical with words, but not with the truth; each one should be explicit and say what *you mean.*

Eventually, you can attempt a complete commentary – start by introducing it. Imagine that your passenger is blind and describe the complete situation, including the car you are driving, road and weather conditions, your location, road number, its direction and mileage to the next town. Then describe the road layout (pavement, dual or single carriageway) speed limits and traffic signs; for example "I'm on a two-lane undivided carriageway restricted to 30 mph". Meanwhile, if there is something more pressing to talk about, bring it in immediately, "Van ahead, checking mirror, braking. No room to go through – holding back until clear". After completing your

Reading the road ahead will aid anticipation.

introduction, lead into talking about the most important situation that is happening, say what it is, how it is going to affect you and what you are going to do about it. Remember that the quality of a commentary is not the quantity of content, but the selection of information that you receive and the correct assessment of its value. Don't "rabbit on" about a lot of nonsense when something vital is about to occur! Base your judgement not only on what you can see, but also what you cannot see and what you can reasonably expect will occur; develop anticipation. Although you cannot be expected to foresee the unreasonable, what is unreasonable to the inexperienced often appears to be quite an ordinary and predictable situation to the more experienced driver.

Plain simple language is by far the best, so use words and phrases as you would in everyday conversation, but don't be repetitive. Where words of one syllable are available use them, and try to avoid multisyllable ones; a common fault and many become tongue tied trying to find the correct pronunciation. Use current and future tenses, never be historical, or hysterical! Do not be afraid of stopping off in mid-sentence if something more demanding presents itself. Keep your voice even and do not be afraid to ad lib. Avoid the monotonous voice, confidential whispers and muttering – try enthusiasm for a change!

The commentary of a professional pilot is calm, calculating, deliberate and decisive. All the facts are there but delivered with brevity. There may be minimal time to communicate information so try to develop a phraseology that is sufficient to describe the situation. However, the context of what you say should be based upon your assessment and anticipation. Any proposed action should be qualified, and all conclusions should be drawn from your observation. For example you might say "There is a garage"; "So what?"! "It has a forecourt which may not be clear"! "So what?" "A car is moving towards the exit, therefore I move away from the exit". That is correct use of observation and related action. A good driving commentator can vary content, volume and rate of delivery to suit any condition. Giving a commentary is an excellent discipline and a great aid to concentration and improved observation. It will extend, broaden and sharpen your observation, allowing an early decision to be made as to course, speed and action, so that you can time control functions to perfection. After all, if you are really concentrating you will be early and will not have to move controls so quickly and then a commentary will not slow you down.

On motorways and rural roads, your observation must be extended into the far distance, working from the foreground, locking on to the relevant hazards. Your eyes should never be still but constantly ranging ahead far and near, over your instruments and mirrors, and always, always moving. Alter your perspective as the density and danger potential of hazards alters or your speed changes. This technique is an invaluable aid in forming your driving plan, enabling you to look, see and assess hazard potential, speed and distance. The actions of other road users can be judged, an appropriate and necessary signal given, with time for others to react before you move. The rest of the driving plan can be implemented with anticipation, consideration and total knowledge of your surroundings. Any warning should always be given early with time for other road users to react and for you to see positive evidence of reaction with time to talk about the hazard before you arrive. If you haven't time, your planning and commentary are too late – timing should be based upon where you sound the horn. If this feature of your plan is too late, the rest of it is too late.

Try to be sure that your phraseology is sufficiently descriptive and brief, for example "Junction on the left, good open view into it and across the field – clear junction right, the mouth is clear but there is potential danger from behind the farm". In rural areas, observation is often not extended sufficiently far forward, or laterally. Houses in the distance may indicate a speed restriction which does not surprise the experienced observer. Tree lines often show the direction of the road over a brow and a farm building in a field should pose the question "Where is the farm entrance?" Use the discipline of commentary to improve your anticipation and 'skill' of observation. In traffic, it is often necessary to break your flow of commentary to mention something new. Be flexible, specific and concentrate on the important hazard first.

Always think of the vulnerable, the elderly pedestrian, cyclist or children and have an accurate sense of speed, be able to 'flow' down the high street, well back from the vehicle in front to see best and be able to drive from one end of a busy high street to the other without braking, if you can. Comment on the position of people, their probable movement and action and be wary all the time of busy folk who glance and do not see or judge speed. This acute sense of observation is why, as your skill develops, you will find that you drive more slowly than many drivers in a busy town, but faster than most in the country. In town, building lines and street lighting will often show the road's direction and the position of junctions, so look over the roof tops and comment way ahead when relevant and possible. Continue talking even when stationary at traffic lights and observe traffic at or near the junction. Use this time to gather information before moving off and remember, it is a good time to check your instruments.

Factual commentary can be broken down into three basic areas; observation, assessment and the driving plan. For example, take a right-hand bend; the assessment is described as "Right-hand bend is acute and blind with no view beyond the apex, I lost it behind the trees and the limit of vision is static". Moving towards the bend, the plan is stated as "On course and holding in to the left, view is still closing. Speed off on approach with light braking. Speed now matches vision and the limited view with gentle power for balance. There is the exit now, increasing speed with view, as I blend power with steering, straightening up. Firm acceleration now to gain the best from the car". Or overtaking, which starts when the 'target' appears in view. "Slow moving box van ahead with a nearside view beyond it." The assessment is described as "I am closing up on the overrun, catch and match situation, I intend to overtake at the first safe opportunity, there is nothing behind". The plan "Speed now matching the van, I am holding off and up to the centre line. I still have a good nearside view. Third gear for best response, still clear behind, moving right, without closing up to confirm view, still clear ahead so overtaking is on. Warning, watch his wheels, move away from the line and I am going with firm acceleration". Clearly a safe manoeuvre with all reasonable precautions applied in good time.

With this simple phraseology you can convey your thoughts and observations to a listener. This self-discipline and training will improve your concentration and it should increase both your and your passengers' confidence, by talking through quite complicated hazards in potentially dangerous situations; perhaps not to be recommended all the time. However, in our experience there is no doubt that an occasional commentary sharpens the quality of driving and helps to develop the easy 'driving flow' of the master motorist that is so rarely seen.

Road signs are, literally, your 'map' of permanent hazards and conditions on the road ahead. Many very good drivers display a considerable weakness in using this vital skill to safe, relaxed driving. One cause of this is the sometimes dangerous familiarity of driving on well known roads. This country has the most comprehensive system of direction signage, numbering, and road marking in the world with clear directions that people often fail to use and understand. Our ordnance survey maps are the best in the world too, so read them well before you start your journey, then you shouldn't have to ask the way.

Your commentary should improve your observation of signs and road markings. As they appear in view give a commentary, developing your observation and forward planning by sighting them as soon as they appear on the horizon. Remember each class of sign – those giving orders, those warning and those informing – has a different shape, handy in snow, fog or contrasting light. Circles give orders – blue circles a positive instruction, red rings or circles a negative one. Triangular signs warn you of an approaching hazard or potential danger and blue rectangles give general information. Green rectangles are used for direction signs on primary traffic routes or major roads. The exception to this general rule is the inverted triangle of the 'Give Way' sign and the octagonal 'Stop' sign that must be obeyed by stopping your front road wheels behind the stop line. When confronted with a 'Stop' sign always use the handbrake as proof of stopping, and always *stop*; very few do! If you cannot see completely, search for view after stopping, by easing straight forwards before you make the decision to go on.

You can stop on an urban clearway but only long enough to set down passengers. Another urban sign is for a bus lane; read the time of the bus lane restriction, many do not use it to keep left outside the time restriction. Note the difference between the blue circle signs, 'turn left', then 'left ahead' and 'keep left' – the white arrow giving the positive instruction. These circular signs are eighteen inches across and are easy to see and read; but are occasionally not obeyed or understood. For example, a speed limit sign can be expected after a 'no stopping end' or a name board approaching a village. This is the time to cut off speed so that you can arrive without braking before the sign. Approaching a national speed limit sign, electrically close your side window, change down for acceleration and although this might appear to be pedantic, don't accelerate (particularly if someone is overtaking) until the sign has disappeared out of the corner of your eye; always remember that the previous restriction is in force until you have passed it.

It really is necessary to see all warning signs as soon as they appear in view, to have time to begin the driving plan as early as possible and to think of the best way to deal with potential hazards. Road works are often inconsistent, you never know what to expect! The information given by the signs is often incomplete and misleading; therefore always proceed with caution. Unfortunately those who place the signs seem to be unaware that misleading information can breed a contempt for them, and a sure way to develop motorists disrespect for the warning is for them to find the road clear. Have you noticed the road man digging on the sign has his wellies on! A level crossing sign seen on the approach to a blind bend or crest could mean the tail of a queue just out of sight, and no approaching traffic could mean the crossing is closed. A single bend warning sign on a winding road should mean that the next bend is sharper than the rest, and a sharp deviation of route sign often means slowing down to second gear speed. A double bend sign will display which bend is first, left or right – many drivers are not aware of this useful warning.

To the automatic transmission driver, a 'series of bends' sign should mean selection of intermediate hold, to give improved control of car speed between bends. An information plate below the sign should tell you how far the bends continue. But is this information noted by the majority of drivers? We doubt it and even fewer would think of linking the distance to the mileometer. Handy to know, particularly if you are following a large vehicle, since it enables you to hang back safely until the distance to the end of the bends is down to one or two tenths, then you can prepare for the first safe overtaking opportunity; linking advanced observation and preparation. Unfortunately, the Department of Transport often give us a double bend repeater every half-mile to drive us all mad – perhaps they don't want us to overtake anyway! Another piece of stupidity is placing a number of warning signs on the same post, so that you have to read from the bottom sign upwards, since that should be the first hazard to deal with! Many motorists don't appear to know the difference between steep hill signs uphill and downhill. Steep hill downwards, read from left to right, means braking to the safe descent speed and changing into the gear to give engine braking to hold you back, taken at the 'Low Gear Now' sign at the top of the hill and not half way down. Heavy lorries should be anticipated to speed up on gentle down grades so don't pass. Climbing a steep hill, however, they can be overtaken on their downchange. By the way, water causes falling rocks in the mountains, so if you see this warning sign look out for ice and rocks, usually on the shadow around the 'back' of the mountain when you start to ascend. A tunnel warning sign should mean slow down, dipped headlights on, sunglasses off.

On motorways and dual carriageways, remember that in the United Kingdom (and on the Continent) traffic often merges from left to right with equal priority, so one should try to use acceleration sense to 'blend and fit' with traffic, without unnecessary braking and avoiding the cross hatching that is there to separate and segregate traffic safely. The dual carriageway ends sign is often mistaken for road narrows both sides and is sometimes not associated with two-way traffic, particularly lethal – so why are they so similar? The end of dual carriage-

Watch for traffic merging from the left so as to blend and flow. If the centre or nearside lane is clear, use it.

way sign is often linked with a double white line system, which can mean no overtaking for both lanes of traffic in addition to cross hatching. The end of the dual carriageway is a daft place to overtake anyway – better at the beginning. So why does it take an age for people to realise the dual carriageway has started?

In order to keep the blue direction signs clear to read at the pace of motorway traffic, it is necessary to limit the number of destinations shown. So when planning a motorway journey, it is essential to check the junction number of your exit first of all. The first sign is usually one mile in advance which gives the number of the road leading from the junction and the exit number in the black panel. The second sign at half a mile, shows the main traffic destination off the motorway. The third, shows all principal destinations ahead with three one hundred yard distance countdown markers to the deceleration lane -so no excuse to miss your exit. A further route direction sign is located in the deceleration lane with a route confirmation sign to follow. On urban motorways, signals are fixed overhead at approximately one thousand yard intervals. Red flashing lights or a St. Andrew's Cross above the lane, indicate stop. If prevented from joining another lane, stop early and well back so that you can still see the signs. You will also come across gantry signs before the exit of a motorway and, if there are downward pointing arrows under the sign, select the lane early for your destination because there will be a reduced number of lanes, perhaps three down to two on the motorway after the junction – look out for the sleepy driver who realises he is being 'turned off' the motorway in the nearside lane and who attempts a late, dangerous lane change to the right. With this junction layout also watch for the driver who tries to make progress by sweeping across, late, from the overtaking lane; a potential conflict with an inattentive driver causing a sharp 'scissors' manoeuvre. Avoid becoming involved by forward planning, early lane selection and adjustment of speed, coupled with proper use of the mirror. When there is no loss of lanes beyond the exit to the motorway, there are separate panels, the one with the upward sloping arrow indicates side road destinations, while the other panel indicates the main motorway destinations. With no loss of lane there is more room at the junction for reasonable progress with forward planning, judgement of speed and distance and acceleration sense.

In an emergency on the motorway, always get your vehicle on to the hard shoulder, angling the front wheels and the car to the left and keep as far away from the traffic lanes as possible; most of the people killed on motorways are killed on the hard shoulder! The emergency orange telephones are at one mile intervals with blue numbered marker posts at every one hundred yards between them; each post marked with an arrow to point the way to walk to the nearest phone. These marker posts have reflective strips which can be seen quite easily in bad weather, and can help you to judge distance. Reflective white studs mark the lanes, the left-hand edge of the motorway has red studs, green for the exit and entrance, with amber studs and a solid white line for the right-hand edge and central reservation, particularly useful in fog – the most potentially lethal time on a motorway.

Contraflows have yellow signs on both sides of the primary carriageway to be sure you see them, even with high sided vehicles in the traffic. (If the authorities closed the nearside lane instead of the one carrying the faster traffic we believe it would cause less accidents in practice). Don't fill space, make it, and keep to the safe following distance and temporary speed limits, which may be mandatory not advisory, in contraflows and fog. On main roads of primary importance, the green rectangular shaped advance direction signs display the next primary traffic destination and route number, some distance before the junction. The 'map' type shows the junction layout, with the most important road shown as a thicker line. Road numbers for later junctions will be read in brackets along the road indicated and for simple junctions or where sign size is limited, destinations can be 'stacked'. For local places, a white sign with a blue border will be closer to the junction, while the junction itself will be identified with a green pointed direction sign. A route confirmation sign confirms the primary route numbers, as a check for you and gives you the distance to the next main town, which you should note on your mileometer.

On fast roads at multi-level junctions, advanced primary route direction signs are seen at half a mile and at the deceleration lane

Road signs tell you what to expect and keep you informed.

with green countdown markers. On busy roads, signs in green are placed above the road similar to those found on motorways with similar downward pointing arrows 'Get in Lane' or 'off set' panels informing you of no reduction in lanes on the primary route. Plain white signs with black borders are used on minor, non-primary routes; they have green panels if the road crosses a primary route and like other advanced direction signs, they can display a warning or prohibition. Complicated, isn't it? You must therefore swot them up if you are not aware of all this information already.

The 'pedestrians on the road ahead' warning sign should mean immediate slowing if your vision is obscured, so that you can stop well within your line of sight. Imagine this situation from a commentary you listen to as a co-driver: "The road surface is good for braking and it's clear and dry. There's a high hedge on the left, the mirror is clear and we are approaching a blind left-hand bend. I can't position towards the centre, as the road is too narrow. There isn't a pavement and someone could be walking towards me, so I must be able to stop. The view is very short so I must also give a light tap on the horn. Low speed round the corner and ready to brake". The elderly couple were there, taking their morning constitutional and an oncoming car prevented us from pulling out. At that speed we could stop gently, without drama. Would you have been able to? But that is not all there is to driving and any silly fool can drive fast enough to be dangerous. You can make more progress to shorten journey time, but drive slowly down busy high streets and narrow single track lanes ;it takes a situation like this to demonstrate why. Speed should be based on what you can see and you reasonably suppose could happen. When you can't see, you should try to give yourself and others time to react. The 'wild animals' warning sign is difficult, as if you travel at five miles an hour, they can run into the side of you. You cannot predict what animals are going to do, and always slow to a crawl for equestrians. Many town drivers never seem to slow sufficiently.

The time of day you are driving is often as important in correct assessment and anticipation of hazards as where you are driving. School, hospital and sometimes factory entrance signs are important in this respect. Hospital signs should mean quietness and low speed, remember ambulances on an emergency do not work office or factory hours! Although only white paint, hazard warnings on the road should give a continuous message, and inform. A line not noticed or often misunderstood is the warning line approaching hazards, relevant in Rules 70 and 91 in The Highway Code but not observed by many. If it is supplemented by continuous road edge line marking, the road is considered to be extra hazardous. There is no waiting alongside the double line sytem and you may only cross the continuous line of the system to turn into or out of an entrance or pass a *stationary* obstruction. You can pass a moving object safely only provided that you do not *touch* the continuous white line with your road wheels. But remember to stay behind a wide slow moving tractor, no matter how slow it is. Overtaking a moving object is a voluntary action, you don't have to do it unless the tractor's wheels stop revolving and then you may choose a safe time to pass. When overtaking always avoid 'clipping' the end of the continuous line, wait until the broken line begins and signal early before it starts, to communicate your intention even though it should be obvious to an impatient driver behind, as it might prevent him committing an offence.

An example of a restriction where the majority see good sense and where it is an offence to park or wait is inside the 'zig-zag' areas on both sides of a pedestrian crossing; it is rare that this parking restriction is abused. Confusion reigns, however, with moving traffic on the approach to the crossing. The impatient will overtake a slow reacting driver. However, it is an offence to pass the front bumper of the lead vehicle before it is clear of the crossing and, if he is the first to arrive, he must be the first to set off otherwise you are overtaking. You can safely slip by the second or subsequent vehicle in another lane on approach, but *not* the first, moving or not – unless it is parked illegally of course.

Intelligent and accurate observation of road signs, both formal ones such as erected by the Highway Authorities and others such as contours, tree lines and hedges are all part of the continuous stream of information the clever and perceptive driver will see, assess and use. It is this stream that should improve your anticipation and apparent 'sixth' sense, if you learn to use it. This keen, intelligent and essential skill of the master driver begins with a complete knowledge of what the signs mean, so study the Highway Code not just when you start driving, but regularly. Use your commentary to ask yourself the meaning of all road signs, formal and informal, then learn from your mistakes so that your knowledge develops with experience. Observation and anticipation are the crucial keys to safe unobtrusive progress; they may also save your car, your life and that of others.

CHAPTER 5

POSITION, TIMING AND SPEED

On the public highway there is far more to think about when cornering than simply the road surface, your line through the corner, speed and stability of your car, and yourself. You must expect other road users and should, naturally, treat them with care and consideration. Furthermore, the 'limit' on a track is your handling skill and your car's performance, whereas on a public road it must always be the ability to stop for other road users, with a margin of safety; in other words with a substantial reserve of skill and handling kept back all the time. Another major difference is that your sight lines on a public highway should be far longer; your focal point is not on the surface immediately ahead as on a track, but through and over the landscape, matching speed with vision and grip. The farther you can extend your vision, the better your anticipation.

Extending vision with speed helps you to plan your course and road position so that your cornering path is exactly placed for each curve. All bends are different in terms of surface, camber, severity, vision and apex. The combinations of vision, grip, cornering and braking will always vary, even from day to day on the same corner. Recognising the 'pattern' and character of a particular stretch of road is the art of an experienced driver. Less experienced drivers invariably try to 'straighten out' bends, often unnecessarily, using the offside of the road in the face of oncoming traffic. You often wonder exactly when or where you can take avoiding action! Such drivers will rarely think of the potential hazards of driving past a 'blind' entrance on the right-hand side of a single carriageway, or one on the left for that matter! Such driving is neither safe nor considerate. Positioning your car towards the centre or offside of the road is simply not safe on narrow 'blind' bends or banked country lanes, unless you have a clear line of sight, and no oncoming traffic. That could be dangerous.

A motorcar is most stable (at any speed), when travelling in a straight line and it is inevitable that some stability will be lost when moving onto a curved path. This loss of stability is directly proportional to the square of any increase in speed. Therefore, the straighter the path of a car through a bend, the greater the stability. If you draw a 'plan' of any bend, the most stable course through that bend is the curve with the greatest radius which can be drawn within the boundaries of the road. Whilst on a track you can always take the line of constant radius, on a public highway safety is the primary criterion; oncoming traffic, pedestrians, cyclists, other road users, and road signs should all influence your path. We discuss the forces on a car and the matter of the correct course through bends later, but this is an important principle of cornering at any speed. On the highway it must be limited by the rules of the road, considerations of absolute safety, and other road users.

Positioning to 'straighten out' bends is only relevant on wider carriageways of over twenty four feet total (both lanes) width. On approach and through the bend there are *three* road positions, you could consider. The normal position in the road could be called a 'safety line position', with an equal margin of safety between the edge of the road and the centre line. There are two alternative positions, a 'reduced safety line' cornering path towards either the nearside edge or the centre line as an extended safety line, depending on whether it is a right or left-hand bend. You should avoid positioning too near the edge of the road as any small mistake or incident, a gust of wind or your brakes pulling to the left, can take you 'off line' onto the gravel at the edge, or even over the edge. As a guide, stay 12-18 inches away, at least.

Very occasionally, approaching a left-hand bend with a clear open view, at speed, there may be the need, for stability, to position yourself over a centre warning line, taking into account The Highway Code, Rule 70. However, before you take such a line ask yourself the following questions. Firstly, will your position be witnessed by an oncoming driver? Secondly, does your speed really warrant the extra stability to be gained? Thirdly, is it necessary to drive a motorcar so fast that you have to increase the effective radius of the curve by encroaching onto the off-side of the road?

The answers to all these questions should be overridden by matters of safety and discretion. Unless you are able to see all the way through an open bend with no 'dead' ground or entrances either side of the road, and there are no oncoming vehicles do *not* take the extended safety line to the off-side of the road; you should not disturb any approaching vehicles. The extent of the safety line will of course depend upon vision, road conditions, road surface and the hazards along each side. In all normal situations, the safety line will be between the nearside and the centre line. If the conditions are 'right' and 'safe' then the extended safety line could be two thirds of the total width of the road. It is very important that you always position your car deliberately, placing it at all times exactly where you want it. Never 'wander' around your half of the road or over the central warning line with only a limited view into a blind bend; your positioning should always be precise and safe.

Three main factors should always govern your entrance, path

Only straighten the curve when vision is clear, with no dead ground or other traffic.

through and exit from any bend. Firstly, you should be well into the *nearside of the road on the exit*. Secondly, your car should be stable throughout the bend and *always be able to remain in to the nearside at the exit*. Thirdly, you should always be *able to stop within the length of your sight line*. Your position relative to the bend should always be governed by safety.

Therefore, at the entrance to a left-hand bend you should only use the two-third road width extended safety line position, if your vision through the whole bend is clear and will remain so, and you can be certain of no roadside hazards. Unless you can, you should always be to the left of the centre warning line into a left-hand bend and into the left in a right-hand bend; your safety margin should be twelve to eighteen inches from the edge in each case. Exiting from a bend your position will be more critical. You should never be witnessed exiting a left-hand bend on the offside of the road. Drifting out near the crown or over the centre or warning lines towards oncoming traffic is very bad driving, or at least displays little consideration towards others. Always be well into the nearside at the exit of a left-hand corner, and able to remain there. Your sense of speed is crucial at the entrance to a bend, as if you enter a left-hand corner too fast on a 'tight' line you will find it difficult to remain on the left of your lane as you exit. When choosing your speed and course for a bend, always take into account safety, view and stability.

Apart from 'vertical' bends, as all hills or crests can be regarded,

A 'blind' gradual left-hander, watch for the house exit and keep to the nearside.

you can categorise 'lateral' bends into six basic types;

1. Open or blind
2. Acute or gradual
3. Left or right

You should use these categories to frame questions as the basis for observing bends as you approach them. The better your observation, the more prepared you will be for hazards, and the greater your opportunity to anticipate what will be 'round the bend'; to stop if necessary. To be able to identify bends quickly and accurately it is essential that you train yourself to 'observe', through gaps in hedges or between buildings, across the landscape; you will be surprised at how much you can identify if you look and concentrate. Sometimes you will be able to see much more as you approach a bend descending a hill than when you arrive, giving you the opportunity to plan your approach.

Many drive along as though they are in a moving 'block' some three times longer and twice as wide as their car, limiting their vision accordingly; they are invariably ill prepared for the hazard. They tend to think about things as they arrive. These are the 'reactive' drivers and just like reactive managers they lack imagination or anticipation. Looking at the road ahead as it disappears into a curve is only of limited, sometimes very limited, value. Use the opportunity offered by the landscape to throw your vision far forward beyond the bend.

You may then be able to 'link up' what you can see before and after the bend to help you to assess its type and severity. Watch for signs of junctions, stationary traffic, pedestrians or other hazards.

The first question to ask yourself is what type of curve is it? An 'open' bend we define as a bend around which the whole of the road surface is visible, giving you the opportunity to 'range' your vision through the bend and along the roadsides. Then, when you have assessed all the potential hazards you should use your view of the road surface to give a sense of position. This sense of position will enable you to select your course, assess the likely grip from tyres, and the effect of camber. This part of your observation is brief, and immediately you have assessed the bend you should 'extend' your view and concentration ahead and far beyond the bend. The faster you travel the longer your sight lines need to be, and the farther 'ahead' of the car your observation; always match speed with available vision. Positioning at the beginning of an 'open curve' is critical, as this will often determine your position through the bend and, very importantly, at the exit. Plan ahead and you can always position for a possible deviation – arrive without anticipation and you are a potential accident.

'Open', gradual bends, or curves we describe as 'engineers' bends, should be taken in the normal safety line position. Only 'flatten' the curve out at the exit for extra stability, if necessary. Vision will be quite clear, and if no traffic is approaching your safety is obvious. Most oncoming drivers will approach such a left-hand bend following the warning line down the centre, so keep away and always watch out for the driver who arrives too fast, trying to overtake or having just overtaken before a bend. Remember that the closing speed of two cars approaching each other at 60 miles per hour is 120 miles per hour, or one hundred and seventy six feet per second. Not a speed to make mistakes at! Always examine the exit of a bend for an escape route, particularly if traffic is ahead of you through the bend. Better to run off than hit an oncoming vehicle. Traffic will convert an 'open' gradual bend into a 'blind' one.

Next, consider an 'open', acute bend, that is a bend which goes through an angle of 90° or less over a fairly short distance. You should approach an acute right-hander, seen to be clear on approach, from the nearside. Then, turn in gradually and a little late from that nearside position, through the bend at a steady speed and throttle, to a late 'apex' around the *back* of the curve, up to, but no more than two thirds of the off-side of the road. This 'apex' gives you the advantage of best camber at the most unstable point of the curve, providing that there are no potential offside dangers, such as 'blind' or concealed entrances. This line will blend into a progressively straight line as you exit, eventually settling in to a normal extended safety line position, with a margin of safety to the left of you, along the nearside of the following straight.

This steering path allows you to apply progressive acceleration after the 'apex', to gain optimum acceleration at the beginning of the straight which is valuable if the straight is longer than the corner, and most are. In slow, out fast and safely should be the rule. The object is stability and exit speed, so your approach speed must be relatively low with a 'late' entry into the bend. Many drivers turn into a bend too early and then have to make a contracting, tightening radius curve with consequent 'tight' exit, with less margin of safety and acceleration. Your approach speed is important, not only for position on the exit, but because the higher your entry speed the greater the tendency for the car to travel straight on at an angle of 90° to the radius drawn from the car's position on the circumference of the curve; put simply with too high an entry speed, you'll fly off at a tangent!

An 'open', acute bend to the left may be approached two-thirds towards the offside of the road, if, and only if, two questions can be answered positively. Firstly, your view of the bend and the straight past the exit, is sufficiently far into the distance for there to be no question of any approaching driver witnessing this approach. Secondly, that because of your speed (not of course due to lack of anticipation!) the extra stability is required. If there are any vehicles approaching you should then take the normal extended safety line, positioning your car eighteen inches inside the centre line. Again a later 'turn in' will bring the steering path through the bend on an expanding radius to a late 'apex', where the inside edge of the road begins to unfold and 'go straight'. The car *must not* drift away from this close nearside position, but be able to remain tight into the left, with a margin of safety; not teetering on the edge of instability towards the centre of the road.

It naturally follows from our definition of 'open' bends, to left or right, gradual or acute, that a 'blind' bend is one around which *the whole of the road surface is not visible*. Even if your vision through a bend is unobstructed, unless you can see the whole of the road surface, we define the bend as 'blind'. 'Dead ground' can be a literal definition unless you treat it with the circumspection it deserves.

As you approach a 'blind', bend you should take all of these factors into account;

> SAFETY to be able to stop well within the distance seen to be clear.
> VIEW since you cannot *know* what is in the blind area.
> STABILITY through the curved path, so that if you have to brake because of a hazard, weight transfer will not 'unstick' the car.

The correct approach to a 'blind', gradual right-hand bend is eighteen inches from the nearside, remembering that any nearside hazard may require a greater *safety* margin. Adverse camber or gravel are hazards that need careful consideration. This nearside position will give the greatest possible margin of safety to approaching traffic, the best possible *view* and line of sight through the bend – so make sure you search for hazards up to the 'limit point' of visibility; not just on the road surface but along each side as well. Finally, *stability* can be increased as far as possible by increasing the radius of the curve through the bend. Remain in the approach position, eighteen inches from the nearside until the view begins to open up. Then, smoothly and gradually straighten out the curve at the exit, to ease the car away from the nearside towards the centre line, just where it begins to straighten; but, only if this position does not 'crowd' oncoming traffic – leave plenty in reserve. Your half of the road opens up first and provided it remains clear of hazards, apply acceleration relative to view. Remember, the presence of traffic will limit your view even further, so stay into the nearside until your view is complete.

By definition, any approach to a 'blind' bend will have limited view, a limited line of sight. Your focal point should be where the left and right verges converge. If, on approach, these two points on the left and on the right move relative to one another the bend is not so acute but rather more gradual.

Entering a 'blind', gradual bend to the left try to keep at least eighteen inches clear left of the centre line on a wide road, *but more if it is narrow*. Extreme positioning for vision is hardly vital if your vision is already very limited and so match speed and vision, with the ability to stop. Always link speed and vision in your mind, and the road must be sufficiently wide for you to position in safety. Remember on narrow country lanes lorries will most certainly 'encroach' over the centre line; other vehicles will unnecessarily. Then all you may see is an early view before they hit you! Even on wide roads consider the intimidation of extreme positioning up to, or even over the line; the latter on a blind bend is irresponsible and possibly dangerous. This is particularly so in fog, when drivers will often use the centre line to orientate themselves!

However, with good visibility on a wide, 'blind', gradual left-hand bend, each additional foot towards the centre line may increase visibility by as much as 100 feet. The constant matching of speed with vision and position, plus acute recognition of potential danger from oncoming traffic will mean that you are always 'safety aware'. Positioning too close to the nearside can be as dangerous as positioning too near the centre line. On narrow country roads particularly, you are inviting disaster from entrances, pedestrians, cyclists or horses; none of which you will see, until too late. So always match speed with vision in such circumstances; be slow and safe. Proper positioning up to but clear of the centre line, as you go through the bend affords safety through maximum view and stability by increasing the radius of curve for the bend. In towns the situation is quite different and you should avoid positioning out to the right. Following vehicles may think you are turning right and 'drive' down the inside!

On a 'blind', gradual left-hander, it is safer to stay out, to the left of but quite clear of the centre, to give yourself maximum sight line and wait for evidence that the bend is finishing, perhaps by a line of trees or buildings running straight. Until the bend opens, contain your speed so that you can stop, comfortably, in your line of sight. Once the bend is seen to be 'opening' and the sight line increasing, you can smoothly and gradually ease the car towards the nearside in a sweeping radius but do not add power yet, as this will increase understeer. Once the car is 'pointing' into the left of the bend towards the straight then increase power for acceleration; if clear and safe to do so. Blend the reduction of steering deflection with increased power if the surface is dry. In the wet, only begin accelerating in a straight line to minimise risk of skidding. Your side will open out last, so always remember, as you point the car towards the nearside on exit, that you must match speed with vision. There may be a slower moving vehicle, a group of cyclists or pedestrians just out of sight. The nearside position on exit also keeps you out of the path of oncoming traffic, or worse still, of someone overtaking late on approach to the bend.

Using this technique properly, linking speed with vision, you always have the *maximum safety*, *best view*, and *greatest car stability*. Entering gradual bends always take up your approach position early for maximum vision; a late course correction may give you an increasingly severe steering path and greater instability. One bend often follows another, so that your exit position from a right-hand bend will result in a smooth and progressive deflection of course to within eighteen inches of the centre line, for correct positioning for the next left-hander.

The correct position for approach to a 'blind', acute right-hand bend is again eighteen inches from the nearside, provided there is no hidden nearside danger. This position gives you the greatest safety margin from oncoming traffic, as well as maximum line of sight. Remain in this nearside position all the way round the bend until you have a complete view of the road surface. As the view opens up return, with minimum steering deflection to the safety line position. If the bend opens out early, then you use an expanding radius curve to an 'apex' towards the centre line. If the corner is narrow and 'blind', consider an audible warning.

On approach to a 'blind', acute left-hand bend take either a safety line or a reduced safety line position, depending on road width. On very narrow, 'blind', acute left-hand bends it may be essential that you do not take any position towards the centre. A margin of safety is vital and correct positioning plus the appropriate speed (which will be very slow if you are correctly matching speed and vision) will provide it. Maintain this nearside position all the way round and be able to stay there on exit. Never forget the possibility of either a stationary or slower moving vehicle on the left, a group of cyclists or pedestrians.

A 'blind', gradual right-hand bend, so position 18 inches from the nearside for view. The telegraph poles and the sign may indicate the exit of the bend. In slow and out fast when vision has opened up.

Match speed with vision at all times, and consider an audible warning; it may be essential.

We have looked at all the main types of bend, without discussing what is an appropriate speed, except in terms of safety, view and stability, and that really is the point. On public roads you will rarely be able to reach the cornering limit of a high performance car with safety, because either the speed limit, vision or safety would prevent it. The conclusion we reach is that the correct position for each type of bend should be clearly laid out in your 'mind's eye' as you approach the bend, using anticipation and observation to evaluate its characteristics. Where you need to reduce speed do so in one, progressive piece of braking using the principles described earlier, and braking, if possible, in a straight line. Always match speed and vision so that you can stop *comfortably* with a margin of safety, within your sight line. After deceleration, make one gear change to the correct gear, and maintain the correct speed through the bend exiting well to the correct side of the road, in a stable condition. The car should be 'balanced fore and aft' as you corner and, accelerate as vision opens up and steering deflection is decreased. Remember, in the wet accelerate in a straight line.

You now have the three 'r's' of driving. In the right place on the road; at the right speed for the conditions and the right gear for the speed – into the bend slowly, out fast and safely. To repeat;

1. Right – position on the road.
2. Right – speed for the conditions.
3. Right – gear for the speed.

Your planning and assessment of a bend starts as soon as it appears in sight. Use the mirrors on approach and having ensured that there is no traffic closing up behind, guide the car onto course for a 'blind' left-hand bend into the extended safety line position, comfortably left of the centre line; don't 'crowd' oncoming traffic. Estimate the correct speed for the corner by looking at the point where the offside and nearside verges intersect at the limit of your sight line. If these two points move in relation to each other, then the bend is not so acute. But look across for any visual gaps in the nearside, however small. Your speed must be matched with vision, so you can stop within your sight line; comfortably. Brake smoothly in a straight line, once, to take off speed, steering with both hands all the time. Select the correct gear for the speed separately and 'cleanly' and then, if the corner is 'blind', narrow and acute, consider an audible warning.

If you are early and anticipate the situation, then these three stages of the plan are completed well before you start steering deflection.

This is just one reason why we are against 'heel and toe-ing' on the public road, it can make you more inclined to leave your braking later. Fine on a track but you need more room to move and more margin for error on public roads. You do not *need* to drive so close to the 'limit', leaving braking late, it simply reduces your 'reserve' of safety which any error on the part of others can erode. On the track you are driving against and with people who you probably know, who are aware of their car's limitations, and who have chosen to race. Margins of safety are different on the road, and not simply for yourself but for all other road users around you. On a track you should race to win, on the road the objective is a margin of safety, consideration for others and unobtrusive progress.

Next commence your steering deflection slowly and gradually. 'Find' the throttle, then gently apply power, *not* to accelerate but to maintain appropriate speed and stable weight distribution for effective side grip. Your line of sight should be as far *through* the bend as possible, not on the road surface any longer than necessary for you to check its condition and your position; matching speed with vision, and grip to stop, if necessary.

Search for the exit of the bend or any clue that the corner is finishing. Look for clues all the time; a line of tree tops that straighten out above your vision line, roofs and chimneys that straighten or telegraph poles, or double lines changing to broken on your side. Then, as your sight line increases and the bend opens up, you gradually and gently apply steering deflection to 'tighten' your line to the left. If your speed is correct you will be able to maintain it, with a little in reserve. You will now be close to the nearside on exit, well away from any oncoming traffic that may encroach over the centre

A 'blind', acute right-hand bend. In slow, out fast, matching speed with available vision.

line. Now, you can apply acceleration as you gradually unwind the slight steering deflection and straighten the car, on the nearside of the road, and with the assistance of any camber. In slow, out fast with maximum safety, and you will always have been able to stop within your line of sight. You have cornered on the curve of reasonable radius, consistent with safety and as your line is straighter, earlier, you can apply acceleration sooner and with safety.

Apply these principles throughout your cornering and you will be observing the correct position, speed, and gear for the bend. You will always be able to stop within your sight line and will have that vital margin of safety. You will always 'place' your car in exactly the right place, never allowing it to wander. You should never drive too fast or too near to the 'limit', but with anticipation and concentration you will ensure you can travel quickly, safely and with consideration. Your position, whether using the safety line or the extended safety line, should always be governed by caution; never assume the road is clear. You may only get one opportunity to be over optimistic! However, always try to 'hang back' and position yourself for view when safe, you will then be ahead of your car and able to anticipate, extending your sight line. Remember that any curve is 'blind' if you cannot see the *whole* of the road surface.

Some of the most deceptive 'bends' are those we have already, albeit briefly, referred to as 'vertical'. Dips or crests, even long wide hills, will affect your car's stability and limit your sight line. Approaching a crest, position yourself to the nearside for safety; you cannot increase your sight line. Reduce your speed before the crest so that you can stop safely within your sight line, then apply acceleration as the view opens up. On narrow country lanes you will certainly need to stop if there is a stationary obstuction just over the crest and an oncoming vehicle. You could also meet slow moving traffic, pedestrians or horses.

Extreme positioning is potentially hazardous and should be avoided when it could create a greater hazard than the hazard itself. For example, when positioning to the nearside for a right-hand bend or for a crest, avoid getting too close to the edge and putting your wheels on to the 'marbles' (gravel at the edge) or the grass verge. Perhaps the most dangerous hazard is the gully that exists on many roads, since that will 'trap' your nearside wheels. If the nearside wheels are on a more slippery surface than the offside wheels, they may well cause the car to spin. Or, if you 'hook up' your nearside wheels in the gully as you turn into the bend, you may spin the car as you try to escape.

Observation, anticipation, and appropriate speed are particularly important on narrow country roads. When negotiating a 'blind' left-hand bend on a road with no pavement, take the safety line position, be slow enough to stop within your line of sight, but without encroaching on the offside; keep left of the warning line. Expect pedestrians, horses, or cyclists, as well as an oncoming vehicle. If all are there at once it will not be sufficient to position out and sound the horn, you must be able to stop!

Emergencies will occur, but driving using these methods will almost eliminate them. Planning, anticipation and a healthy pessimism are essential to safe, swift motoring. Car control at the limits of grip and braking are to be studiously avoided on the public highway but, if an emergency does occur you should try to have the mental and physical control to take the *minimum* amount of evasive action. This requires good judgement and a great deal of experience, preferably at the 'limit' on track and skid pan. If you drive with anticipation, consideration and an appropriate sense of speed and position you should never need to react harshly with the controls; even in an emergency you should be calmly in control. Unfortunately, few drivers ever have the experience of skid pan or track driving to develop their car control skills, not to be a 'racer' but simply to be able to survive under all circumstances.

In our view, to drive on a circuit or skid pan without qualified instruction, even if you are alone, is irresponsible. Porsche Customer Driving Days, organised by Porsche Cars Great Britain Limited on behalf of its Official Porsche Centres, are an excellent opportunity for you to gain experience of car control under supervision and with expert guidance. We strongly recommend that you take advantage of such opportunities, regularly. The more often you use such practice the more your skill and reflexes become conditioned, your concentration improves and you will be a safer motorist. However, always remember that racing lines and appropriate safety lines for the public road are very different. Racing lines are potentially lethal if used on the public road, particularly on 'blind' two-lane roads.

Whichever Porsche you drive you have the pleasure of one of the world's great motorcars. Not for you the dull stolid front-wheel drive, but the joy of a beautifully 'balanced' chassis, with excellent handling and power. The skill you will acquire as you gain experience and knowledge should be held in 'reserve' for emergencies. Emergencies you will certainly avoid if you drive by our principles of anticipation, planning and consideration. An accident is usually a build up of circumstances, perhaps only one of which need actually be out of your control. The correct professional training, learned experience and, a car that will also have something in 'reserve', as well as the driver, makes for safer, surer and more enjoyable 'motoring'.

CHAPTER 6

FORCES ON THE CAR

There are many factors, apart from you, the driver, that influence the manner in which a car will behave when it is in motion. We have already agreed that a stationary car (safely parked of course), is no problem to anyone! Knowledge of these forces, of a car's basic design characteristics and how, together, they affect the car's handling, will help your understanding of what goes on between you, the controls, the tyres and the road. The Ministry of Transport Driving Test requires no practical knowledge of skidding, motorway driving or car control theory, and yet driver error is a major factor in at least 90% of accidents. All drivers should be aware of the forces acting upon their car, how they can be controlled, and perhaps more importantly, how to stop a car going 'out of control'.

Many factors affect a car's stability, and influence the relationship between road and tyres, including the driver. But, apart from the driver, by far the most important factors are the tyres and speed. Speed is the major cause of skidding, more important than tyres, road surface or anything inherent in a car's design. The inexperienced driver without any experience of skidding, will react 'naturally' in a skid, handle the controls harshly and fail to 'blend' the various forces acting on the car to obtain the best grip between road and tyres. He or she will often overreact through a lack of sympathy and often cause a skid, or worse!. Drive smoothly, with an appropriate sense of speed at all times, and you will rarely skid.

Two simple examples will demonstrate the crucial effect of speed in the practice of car control. Take several flat stones of equal weight, say two or three inches across. Remember the game of 'skimming the stones' you used to play at the seaside? Find a flat piece of water and throw the stones, at different speeds across the water, so that they 'bounce' off the surface. The faster you throw them the more they bounce, providing the angle at which they strike the water is right. The weight of stone, air and water resistance are all constant; only speed varies. That is exactly what a tyre does on the road. If speed is too great or you apply harsh acceleration, braking, or steering, at the wrong time, you will just make matters worse.

Weight in motion, desires to travel in a straight line. When you hang a weight on the end of a piece of string and swing it round holding one end tightly in your hand, Newton's laws of physics will tell you that 'for every action there is an equal and opposite reaction'. The pull you generate on the string, prevents it from flying out, but the faster you rotate it the stronger is the pull it exerts on your hand. Place several loose objects on the left-hand side of the rear seat of your car, take the car round a left-hand bend and all the loose objects will fly over to the right; it will happen the other way too!

A car cornering and a weight on the end of a piece of string will both tend to fly-off at a tangent to the curved path they are travelling, at right angles to the line of radius; this is known as centrifugal force. The forces acting on your car in motion are wind resistance, friction between tyres and road and the moments of inertia around the vehicle's centre of gravity. Understanding how these basic forces work will help you to know what is happening and why. These forces are not of equal importance so one we will ignore is wind resistance. The most important forces are;

1. Centrifugal force C.
2. Sideways force S or adhesion.
3. Driving force M.

Centrifugal force always acts about the car's centre of gravity 'g', and is therefore always above ground level. It forces the car out of the cornering path at a tangent, and tilts the car onto the outer wheels. Sideways force, or grip, and driving force both act at ground level, where tyres and road meet, and will exactly balance the centrifugal force. These two forces can never be greater than the adhesion force between tyres and road. Centrifugal force C is trying to take the car off the road at a tangent to its cornering path, whilst S and M are trying to keep it there.

Let us consider the interaction of these forces in more detail. M, the driving force, is a simple function of power applied to the driving wheels, front, rear, or all four, and gives velocity V, through the corner. The side force is the adhesion 'f', between the tyres and the road, which is also a function of the car's weight W. The radius of the curve 'x','y' is 'r', between 'g', the car's centre of gravity and O, the centre of the arc described by the car's cornering path. In the diagram overleaf the two turning moments acting in the plane of the car's motion through 'g' are C, and the horizontal turning moment T, tending to spin the car.

In diagram 6/1, the car is describing an arc 'x','y' with radius 'r', from the centre O. The centrifugal force C acts through 'g', the centre of gravity of the car, and would pull the car off the curve, were it not for the adhesion force of the tyres. This adhesion force 'f' is a simple function of tyres, road surface and the car's weight. The balance of these forces will determine the speed at which a particular car can negotiate a corner, and also help to explain the handling characteristics of different cars. The shorter the radius of the

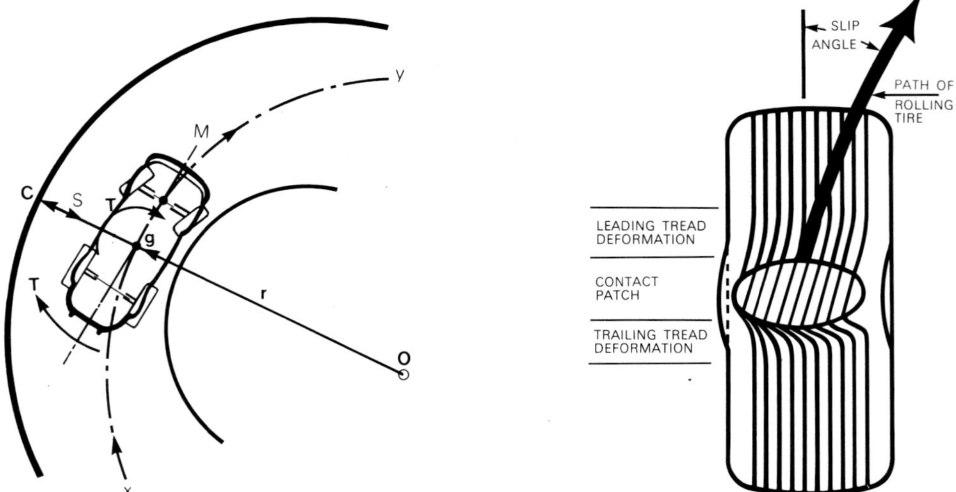

6/1
(Left) The main forces acting on a car when cornering.

6/2
(Right) The distortion of a tyre tread under cornering, viewed from below.

curve, the larger the forces of adhesion have to be for a particular car to stay on the road at a given speed. The formula for the maximum theoretical speed, on a perfectly flat road is :

Equation 1. $C = \frac{W}{g} \times \frac{V^2}{r}$

where; C, is the centrifugal force which would force the car to the outside of the curve.
V, is the velocity of the car, squared.
W, the weight of the car which when multiplied by 'f', the coefficient of adhesion of the tyres to the road, is the equal and opposite force to C if the car is to stay on its cornering path.
'g', is the force of gravity, 32.2ft/sec^2.

From Equation 1, since C = Wf, simple substitution gives us,

$$Wf = \frac{W}{g} \times \frac{V^2}{r}$$

from which $V^2 = f \times g \times r$

The critical importance of your speed, tyres, the road surface and the radius of the arc you describe through the corner, are obvious. Note that the sideways force (f × g × r) is a function of *the square of the speed V*. As you increase speed through a corner, the sideways force has to increase with the square of that speed.

Tyres are the most important factor, as they generate cornering power. Your cornering car is held on its line by the four patches of tyre rubber in contact with the road; a total area approximately twice the size of this page! How does a tyre generate cornering power? In order for your car to change direction the tyres have to develop friction, at any speed or radius of curvature; they have to overcome the centrifugal force. As the steering wheels turn, the tyre tread in direct contact with the road is distorted; because it is elastic it will not turn as far as the wheel rim does. The path that the tyre follows on the road surface is different from the direction the wheel rim is pointing; this difference is the 'slip angle'. In diagram 6/2, we show how a cornering tyre will appear looking up from the road surface. You can see the distortion of the tread, and the slip angle.

In diagram 6/3A the car is travelling at moderate speed, and the slip angle of the front tyre is apparent. The rear wheels will follow a path slightly inside (on a smaller radius) the front wheels. This is because the rear wheels are always 'straight' and will initially tend to drive inside the front wheels. At very low speeds the slip angle will be small, but the coefficient of friction that a tyre needs to generate to 'stick' your car onto a cornering path, increases as the square of the speed (or with the severity of the corner). This forces the driver to steer into the corner, increasing the slip angle and generating greater grip. The front wheels have a greater slip angle (they are pointing inside the turning path) than the rear wheels; the car is understeering.

Unsettle the car, and the rear driving wheels will tend to lose grip against the increasing polar moment of inertia, and describe a larger arc. They can only generate higher sideways force 'f' by increasing their slip angle, and will start to point the whole car into the corner. Correction has to be applied to the front, non-driving, steering wheels and steering lock is reduced; reducing the slip angle of the front wheels. In diagram 6/3B the car is shown drifting in an oversteering position, the slip angle of the rear wheels is greater than the slip angle of the front wheels.

The outside front steering wheel has slightly different forces acting upon it because 'g', the car's centre of gravity, is above ground whereas the point of all the other forces is on the ground, at point of contact with the road. Even on a Formula 1 Grand Prix car, 'g'

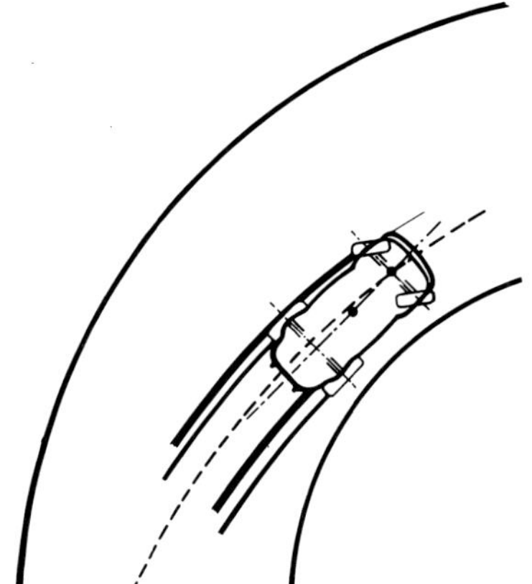

6/3A
(Left) An understeering car.

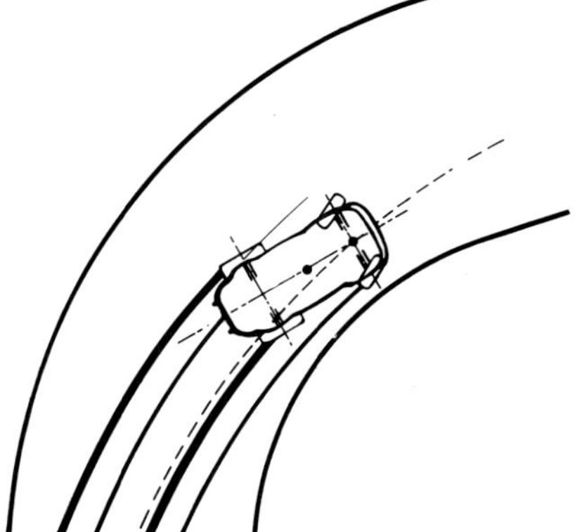

6/3B
(Right) An oversteering car.

is not on the ground and any cornering force will cause a car to tilt outwards, there will be a turning moment tilting the car outwards.

The handling characteristics of a car, when and how it changes from under to oversteer, are a function of how the suspension and design layouts transmit those forces to the road. Imagine you have a weight in your hand, the flat sort that weight-lifters use. Rotate your hand back and forth and you can stop it easily; the weight is evenly distributed around the axis of rotation. Now take two different weights, equal in total to the first, and stick them onto a short lifting bar. Now try to rotate the bar back and forth as you did before. It is difficult to get moving in one direction and more difficult to stop it and rotate in the opposite direction. The former has a low 'polar moment of inertia', with total weight concentrated about the centre of gravity. The latter has a high 'polar moment of inertia', with the weight away from the centre of gravity.

There are many ways that a designer can alter, or build in, under or oversteering characteristics, and we will not dwell on them here. However, from the road driver's point of view, these are two important elements of a car's handling characteristics. Thus, when a car is oversteering, you have to apply a steering correction on the front wheels to decrease the car's natural turning radius. Without this steering correction the car will oversteer itself into the inside of a corner. When a car is understeering, you have to apply a steering correction to 'increase' the car's natural turning radius. Without steering correction the car will follow a curve of greater radius than intended, and understeer out of the corner.

The most desirable road cars are those with a high moment of inertia and equal weight distribution around the front and rear axles. This wide distribution of a car's weight, gives a slow and predictable response that can be easily 'felt', and corrected. Mid-engined cars have a low polar moment of inertia, they are very manoeuvrable but 'twitchy', and once they start to spin, tend to go very quickly and require great skill to correct. Suspension designers usually 'build in' initial slip at the front wheels, both for stability and to give a warning of slip angle to the driver's hands; it is felt as increased lightness on the steering wheel. If the change from this initial understeer to oversteer is gradual and progressive, then a car is said to handle well. The driver has plenty of warning before oversteer develops and can therefore correct it.

But weight distribution affects not only the lateral turning moment around 'g', but also the vertical moment. In cars with a rear weight bias, as you accelerate the weight moves towards the already heavier rear and creates downward pressure on the rear wheels. As you decelerate the reverse happens, and weight is thrown forward on to the front wheels. Under acceleration, a car with rear weight bias has better traction at the rear wheels and under deceleration you will feel less loss of adhesion than in a car with a front weight bias.

(Opposite above) The current Porsche 911 Carrera is an inherently understeering car.

(Opposite below) To make a Porsche 911 oversteer you have to unsettle the car in a corner.

SEARCHING FOR LIMITATIONS

The best way to experience the handling characteristics of your car is with the guidance of an expert professional co-driver, and the safest place is a clear race track or disused airfield, with large run-off areas; *never the public road.* Let us imagine you have found such a person and place, how do you go about testing your car's handling?

An ideal place to start, is a large flat surface with even grip and a marked circle of one hundred feet radius. Remember, that in all of the following manoeuvres your 'touch' on the controls must be light, smooth, but positive; not jerky or harsh. Start at a relatively slow speed, say twenty miles per hour. Turn into the marked curve at a constant speed. Deflect the steering wheel slowly and positively, until you are running round the circle. You will feel the sideways force on the car by being pushed towards the outside of your seat. Never 'hang on' to the steering wheel – you will not only lose 'feel' but you may inadvertently alter your cornering radius. As you progressively and gently increase speed you will change the weight distribution around 'g', and the polar moments of inertia!

Taking your car round the circular path, increase speed by small amounts, circulating at slightly higher constant speeds, with minimum power and steering each time. Remember the square rule, if you increase speed, the cornering forces must increase as the square of that speed. You control the speed through the accelerator and, if the car is understeering, back off the accelerator gently to reduce the understeer, causing the car to 'tuck in' to the bend. But make any adjustments gently; 'feel' the limits of the car. Next, lift off the throttle quickly and feel the rear wheels start to 'slip'. Then you will have to 'un-wind' the steering, and steer into the skid and around the circle, to keep the car on the marked path. Quickly lift off the power, recorrect and straighten the steering as the car recovers.

Always remembering that cornering force varies with the 'square' of your speed, you need only make alterations in speed. Don't rush this process of learning, and don't be afraid to spin; the large run-off area and the fact that you are approaching each stage gently make it safe. But you must 'feel' every change and shift in the balance of the car, very difficult if you have never experienced the motion of skidding before. Yet that is exactly what happens to most drivers before crashing; they never experience a skid, until it is too late! Better, safer, and cheaper to have an expert demonstrate the skills required on a skid pan, at a slow, safe, speed in an 'old banger'; but that may dent the ego of the driver who thinks that he or she is 'good'. If difficulty is experienced in basic car control in such ideal conditions, then the ability to 'cope' in an emergency simply does not exist!

Next, start on a clear skid road by driving backwards, fast; yes backwards! Feel how unstable most front-engined cars are in this situation, with front engine weight at the wrong end for stability,

(Opposite above) Under acceleration a car 'squats', giving greater traction on the rear wheels.

(Opposite below) Under braking weight transfers forward and the front wheels may lock-up first.

and opposite 'castor' angle with the steering in reverse. Reversing *quickly*, steer quickly onto left lock, disengage the clutch as the front of the car goes through 90°, take second gear and zero the steering gradually, for straight 180° ahead. Try this, 'front end throw-over', to both the left and the right, until you are instinctively aware of which way to steer when you are spinning backwards. It is an excellent exercise for improving your orientation in a spin. Without such practice you will, literally, not know where you are.

Now try a 180° spin forwards. At a brisk pace (faster on a better surface) deflect the steering onto right lock first, to create sideways force, then declutch and lock the back wheels with the handbrake; if it is 'man' enough for the job. Then, as the front of the car goes through 90°, release the handbrake and start to centralise the steering as the car straightens. Run straight backwards, slow down a bit with the brakes and double-declutch into reverse to keep full control. With rear-wheel drive you can try a full power wheel spin to break traction, after deflecting the steering onto right lock, but only on a very slippery surface. Then, off the power, declutch at 90° and follow the procedure above. Now, full speed in reverse and try the 'front end throw' again. Great fun, and taking it in stages will help to develop your car control, thoroughly.

Now for the 'triple' – complete a 360° spin to the right. Start at speed from the left side of the skid road because with the wheels turning, it will end up over to the right side of the road. Steer right lock, (declutch), pull hard on the handbrake to break adhesion of the rear wheels, and as the front comes through 90° steer quickly onto left lock. Then, gradually start taking off steering and engage a lower gear (because it slows you down), continue to centre the steering as you come to 360°, and you have done it! For 360° left, start from the extreme right of the road. Steer left, on the handbrake declutch at 90°, turn quickly onto right lock and (remember don't panic and lock the wheels with the brakes, or it may not continue to spin round) steer right, change down and zero the wheel – some correction may be required to keep the car straight on completion.

After lots of practice, you will be able to place the car exactly – it really is very predictable. The car will not turn over unless it hits a curb, even on a dry road. The exception is if the tyre pressures are too low for the grip, racing a saloon car fitted with high grip tyres, or if you attempt the 360° too timidly, at too low a road speed. *That's why it is only safe to learn spin technique in stages on a slippery skid pan with plenty of run off areas, and expert tuition.*

What use is all this? Well, if you spin on the road, you will keep 'working' to place the car into the part of the road you wish to and recover control to carry on driving. Spinning or losing control has to be considered an accident even if you do not hit anything as it is potentially lethal if you do. Practising car control skills of this kind has a salutory effect on all drivers we have known who have undertaken them. Firstly, it gives a very precise sense of what your car will do and when. Secondly, you realise how quickly the transitions from understeer through oversteer to a spin can happen. Thirdly, you develop your handling skills so that you *know* 'where' the car is and use steering, brakes and power to control the car properly.

When you first drive a car on a circuit it is safer and preferable to take it *easy* at first; concentrate on lapping *smoothly and precisely*, just 'tracking' round. Then build up speed each lap, gradually approaching *the limit*, but still leaving that essential margin of safety at the exit of the racing line at each corner, say two to three feet. This is the safest approach although it takes more time, but time and space are precisely what you use a track and skid pan for. As with the exercises on a skid pan, we recommend that any practice of the type we are describing, should be done on a near empty track; preferably with professional instruction at first.

After skid pan training, practice on a wet track is an excellent means of getting the 'feel' of your car at *the limit*. The basic cause of skidding is always the same, wet or dry; the kinetic energy of the car overcoming the cornering forces of the tyres. Speed or brakes are the control cause, and you must therefore react and remove the cause. But always use only the minimum control change necessary.

It is very important to assess speed for bends and curves precisely, have confidence to corner at the speed you have chosen with a steady throttle and one steering angle, progressively and slowly applied on approach. Circle the car with the minimum adjustment of steering and power to corner around the constant radius of our painted circle and find the limit of your car's road holding. Is the 'limit' of the car, your own, or the 'limit' of the car's ability? If you spin safely, it will not matter, the car will not turn over unless the wheels touch an obstruction. In the search for the 'limits' of yourself or your car, you will spin now and then – if you don't you are not trying! Trying hard on a consistent surface with a constant radius and you will see, very clearly, how speed is the major factor. A *slight* increase in speed has a dramatic effect. Once the 'limit' has been achieved consistently, you can only go faster by steering a wider path on an expanding radius,outside your circle; accelerating on an expanding radius. Take a contracting radius when turning into the bend, then a constant radius at a steady speed and 'drift' – perhaps with the car pointing as much as fifteen degrees into the corner – and finally an expanding radius on exit. The earlier the actual increase in exit speed occurs, the faster the car will begin the following straight and the advantage is maintained all down the straight. This technique demands great skill and practice to stay under control and the circuit is the only acceptable place for this kind of fast driving. Knowledge and skill developed should be kept in reserve for emergencies or, as a challenge, for fun in motor sport, not the public road.

Let us summarise; briefly. The designer has given the car the best road holding with the engine just pulling the weight of the car, not accelerating, but overcoming the friction losses. The accelerator not only controls speed, but also the longitudinal tilt of the car on its suspension, front to rear; remember acceleration transfers weight. This weight transfer presses the tyres into firmer contact with the road surface and increases the area of the tyre contact, giving more grip at the back tyres. A good driver will 'balance' all the forces by controlling speed and tilt of the car with 'squat' under power or 'dive' when lifting off the throttle. Once equilibrium is created, don't suddenly lift off the throttle or jerk the steering. Not only will you upset the balance of speed and line through the bend but you will transfer weight slightly from the rear of the car to the front. When a car is understeering, back off the power and understeer is reduced, causing the car to turn into the bend; 'lift off, tuck in!'. When a car is oversteering reducing the power will 'un-weight' the rear wheels slightly. Skiers will understand what we mean by 'un-weighting'; lifting the tail of the skis to turn. But, if speed is the cause of the instability you must remove it, by easing off the power, *early*.

On the highway, there must always be plenty in reserve, and a safe sense of speed is essential; you must be able to 'read' the road surface to avoid trouble. Judging the correct approach speed for a hazard is not easy, many fail to notice any change in grip and assessment of speed and distance is often faulty. This judgement will be improved by circuit driving and 'feeling' the limits of your car yourself.

Design Layouts and Handling

You will have understood, from what we have written, that the weight distribution of a motorcar has a fundamental influence upon handling. The advantages and disadvantages of various layouts have been the subject of continuous debate, ever since the motorcar was invented. The racing car layout which Professor Dr. Ferdinand Porsche devised for the wonderful Auto-Union P-Wagen in 1932 is now, with a number of modern amendments, the standard layout for a Grand Prix car. However, the modern Grand Prix car does not have to contend with the weight variations of passengers or luggage. The front-wheel drive, front-engined layout of the modern era, usually has heavy understeering characteristics, and many designers do not consider it is suitable for high performance cars.

Porsche has undertaken a great deal of research into the different characteristics of various layouts. Each layout has certain advantages and disadvantages, depending upon whether the vehicle is laden or unladen. The Porsche Transaxle design offers, in our view, major advantages for a road car. The almost perfect 49% front/51% rear weight distribution of an unladen car, plus the wide spread of the weight, gives a high polar moment of inertia. This gives both predict-

Weight distribution of various types of design

Conventional Design

−Unladen: Little weight on rear axle = bad traction of driving wheels (especially in winter and on bad roads).

Rear Engine

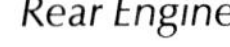

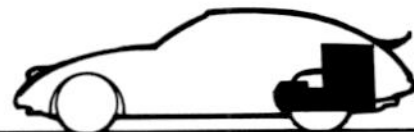

+Unladen: Heavy weight on rear axle good traction of driving wheels.

Front-Wheel Drive

+Unladen: Heavy weight on front axle = good traction of driving wheels.

Transaxle Design

+Unladen: Heavy weight on rear axle (through transmission in rear), thus good traction of driving wheels.

+Laden: Heavy weight on rear axle = good traction of driving wheels.
+Large, variable luggage compartment.

+Laden: Heavy weight on rear axle good traction of driving wheels.
−Split luggage compartment.

−Laden: Little weight on front axle = bad traction of driving wheels (winter and bad roads).
+Large, variable luggage compartment.

+Laden: Heavy weight on rear axle (through transmission in rear + part of load), thus good traction of driving wheels.
+Large, variable luggage compartment.

Design configurations.

able handling characteristics, as well as a slowness of reaction to lateral rotational forces. This neutral handling and the gradual change from understeer to oversteer, makes them very desirable road cars.

Rear-engined cars have other advantages – but do require rather more sensitive handling. They provide excellent rear-wheel traction under acceleration and deceleration, and can 'cope' with a greater amount of power. The main point to remember is that it is the transition from understeer into oversteer that is critical in assessing a car's handling. Rear-engined Porsches have extraordinarily good cornering ability and modern tyres are very predictable. Nevertheless, the transition from understeer to oversteer is not quite as gradual as with the Transaxle cars; they need a little more 'respect'.

Four-wheel drive is now a feature of many modern road-going cars, including the 'ultimate' Porsche, the 959. The driveline affects a car's handling characteristics and the four-wheel drive concept produces very different characteristics. The power or driving force M is transmitted through all four wheels, equally or through a differential or variable torque drive system, depending upon the particular model. This gives two main handling features. Firstly, whether the car will under or oversteer depends upon the power distribution to all four wheels and secondly, as the drive force is split between all four wheels the driver can have less flexibility, particularly with a fixed torque split. Nevertheless four-wheel drive gives outstanding traction and with a flexible torque split offers all of the advantages of single axle layouts, with few of the disadvantages. Speed is controlled more effectively, through all four wheels.

Drive configuration will never reduce the crucial importance of good tyres, correctly inflated. Always use correct tyre pressures and follow the recommendation of the Porsche engineers at Weissach – never use tyres with less than 3mm tread depth (the U.K. 'legal' limit is 1mm). Wrong tyre pressures, excess tread wear, or using tyres that do not match the car manufacturer's specification, will alter a car's handling characteristics, probably disasterously. Then, the only predictable feature will be unpredictability.

Braking on the Limit

The nastiest experience you can have when motoring is to have to stop quickly in an emergency. The most natural reaction (like a drowning man clutching at straw), is to 'bang' the brake pedal hard and fast; you react but do not think. The car then skids straight on into danger, down the line of momentum, with no steering control and all four wheels locked. The impact is in proportion to mass and the square of your speed! With the tyre contact patch *sliding* across the surface, it matters little which way you angle the steering, the car will not respond. It will stop pretty fast on a dry road, or if the front wheels are angled in soft snow or gravel; as that forms a wedge and helps you to slow. However, release the brakes with an angle of steering already applied and the instant the wheels roll, the car will react, and 'dodge' around the hazard. Easier said than done, we know, to keep presence of mind in an emergency. Practice on the skid road will develop your confidence in this technique or at one of the Porsche Customer Driving Days organised by Porsche Cars Great Britain Limited, but don't try it at high speed, or the car will spin like a top. This technique *must never* be practised on the public highway, but it may save your life in an emergency.

On all road surface conditions, wet or dry, ice or snow, the best braking grip occurs *just* before the wheels stop rotating. In a practical sense, the only way you can prove you have the best braking grip in an emergency is to brake harder and more firmly until the wheels do lock. You do this with a quick and progressively harder push – don't 'bang' the pedal. The moment you feel that slight bump, caused by the loss of rotating inertia of the wheels stopping, release *just* enough pressure to get them turning again. On ice you will have to release the brakes completely, to allow the low adhesion to turn the wheels again. But, on a dry road, keep most of the braking pressure on, as long as you are sure you have released sufficient pressure for the wheels to rotate again. Don't fully release the brakes, you won't stop doing that! Lock all four wheels, not just the front two, then release, then increase pressure to lock them again, then slightly relax pressure – increase – relax – increase – relax, until the car stops.

Cadence braking takes great skill and practice on different surfaces and cars – they are all different. As with most advanced techniques, there is far more to it than one first supposes. Although some master drivers may be able to stop in a shorter distance than anti-lock braking systems (known as ABS) on certain surfaces, will they do it when they are tired, or on a 'dirty' wet night at three in the morning? We doubt it. Don't try to cadence brake as hard and as fast as you can by 'banging' the pedal, or transferring weight by timing it with suspension spring phase; and remember to disengage the clutch! Keep the nose down and fluctuate pressure with as much sensitivity as possible, locking and partially releasing all four wheels skilfully. The advantage of cadence braking is that you not only stop as quickly as possible, but you maintain your steering control and therefore stability. You can steer slightly towards safety at the same time as stopping to avoid the accident. It is perfectly possible to set a car up into a 'tail out' oversteering slide on the skid road; keep balance, stability, and control and cadence brake to a stop. You can keep a car balanced in a bend with brakes and steering, just as you can with the accelerator and steering.

Correct, skilful and efficient cadence braking is frequently misun-

The effect of ABS brakes.

derstood, even by the most experienced drivers. When anti-lock brakes are fitted to your car, they have to be used properly. John Lyon had one person say they were not fitted to his car – they were – he was simply not braking hard enough to get them to pulsate. Another driver wondered what the pulsating was when he applied the brakes! There is no substitute for practical driving instruction by a master in the end. The value of ABS braking systems can be properly and safely demonstrated on a skid road, as you can see from the sequence of photographs elsewhere.

To demonstrate the advantages of ABS against locking the wheels and then cadence braking to a stop, we set up a test at Brands Hatch; on a day when the circuit was free. The results are shown in the sequence of photographs. John Lyon could not beat the stopping distance of the Porsche 944 Turbo with Sport Equipment the only car in the group fitted with ABS; could anyone?

One of the worst situations is to have to stop quickly when you have one set of wheels on the dry and the other on a slippery surface; ABS will work in this situation too! To demonstrate this try it on a track as we did. Drive quickly with one side on a wet surface and the other on the dry; brake hard and you will feel the wheels on the slippery surface locking and unlocking with rapid pulsation. It works better than any driver, keeping the car straight whilst stopping very quickly. Try it without ABS brakes and you will spin like a top! Even in this extraordinary situation, you can steer towards safety and avoid the accident. The great advantage of anti-lock brakes is the ability to alter course, whilst still stopping quickly. ABS works far better with the clutch down; disconnecting the drive will let you feel the system, as you will have to when cadence braking, to prevent the engine stalling. If the engine stalls you won't be able to restart it again until you stop. *So clutch down to free the wheels from the engine and gearbox but only when intending to lock the brakes.* In an emergency you could stall on the initial braking, but it matters little, just keep the drive free while you 'cadence' to a stop. But you must actually lock all four wheels otherwise the stop will not be so effective.

Ice, Snow and Water

Coefficients of Adhesion between Tyre and Road

Type of Surface Dressing	*Dry*	*Wet*
Tarmacadam	.8	.55
Rough concrete	.8	.55
Smooth asphalt	.7	.4
Cobblestones (pavé)	.6	.4
Wood blocks	.6	.3
Snow	0	.2 to .15
Ice	0	.1 to .05

These coefficients are approximate and may vary by $\pm10\%$, depending on road conditions, the design of tyre treads, and the type of 'mix' used in their construction. On wet roads, for example, the coefficients increase by .1 with well treaded tyres, and decrease sharply if they are smooth. They also diminish slightly with a rise in speed.

Having the experience to sense an icy surface is vital to avoid trouble. The most difficult hazard to spot is black ice at night, although you can sometimes see a reflecting twinkle of moonlight or headlight reflection but only by taking *great care when you know it is freezing*, are you going to stop a potential accident. After all, it is evident when you open your front door and walk out on the ice to open your car, is it not? Two other warnings of ice are a feeling of lightness in the steering created by the castor angle and offset centre point steering geometry, and the silence of wheels running over ice – still nasty if you realise too late to avoid loss of control. Quite frankly, if you are travelling too fast on ice, no skid road training is going to save you if an emergency happens. Nor will four-wheel drive or anti-lock brakes, kinetic energy will be the master. Ice has less than one tenth the grip of wet tarmac, and if you fail to notice an icy surface in the distance and you are travelling at fifty miles per hour, there is probably no way you can take any kind of corner. The car will go straight on down the line of momentum of kinetic energy, you may even feel a sense of acceleration with so little grip to oppose it; and imagine, ten times the recommended highway code braking distance to stop!

In this instance, momentum acting in a straight line on approach, or any centrifugal force created by trying to corner, will act mostly where the engine weight is. When cornering, these forces will push on the centre of the mass and, with hardly any opposing force created by the grip of tyres on the road, the car will fly off at a tangent, either nose first or rotate tail first depending upon weight distribution. With the engine over the driven wheels, there is, however, the advantage of good traction; good for ascending hills. But rear-engined cars tend to lock their front wheels when braking on ice, and vice versa with front-wheel drive. There is no skid worse than the front sliding straight on.

If you experience a rear-wheel skid with a rear-wheel drive car, ease your foot off the power, keep the front wheels pointing down the path you wish to follow and 'turn into the skid'. Recorrect, that is straighten, the wheels when the car recovers from the skid (if you are lucky), then 'tickle' the power again, only when you are completely back under control. Better than all of these techniques however is to observe, anticipate and see the ice patch in plenty of time, and brake on the good surface before you get to it. Only put the clutch down at walking pace on sheet ice, or when you wish to lock the driven wheels when you are cadence braking. Don't use the gearbox to slow down, or worse, 'clutch drag' from an un-matched change down, as you will cause a skid with the two driven wheels – okay with four-wheel drive. Delicately brush the brakes first of all and cadence, on, off, on the brakes, if necessary. Keep your eye on the mirror, be very delicate with the controls, stop gradually and progressively and watch approaching traffic like a hawk for anyone who looks out of control or likely to cross your path, particularly articulated lorries, they can kill you if they 'jack-knife'. Lastly, remember tyres running over ice are quiet – they make *no* noise!

Packed snow is tricky when it freezes, having little more grip than ice. Locked wheels, with full steering lock applied to form a wedge in soft snow are effective to stop at low speed, but cadence braking with all four wheels is safest at higher velocities. In slides created by excessive speed or power oversteer with rear-wheel drive, lift off the power instantly you sense the skid is out of control – most people are too late reacting and lose control. Turn into the skid or 'point' down the steering path you wish to follow, re-correct to straighten the steering when the tail slide recovers. There is just *no* substitute for practical training on a skid pan – government departments take notice!

Remember that all you are trying to do is lose speed and keep the front wheels pointing the way you wish to go, so look where you want to go and not at the thing you might hit, and only go back on the power when you are completely under control. 'Fish tailing' down the road is caused by not removing the power completely or late re-correction of the steering. (One of the reasons, we believe, the police say that most skidding accidents happen on a straight road!). If the front slides away in front or four-wheel drive through excessive speed or power, remove the cause and keep the angle of steering applied to form a wedge of snow to help you to slow down. Lifting off the power and using engine compression to slow the car through the driven wheels and tyres that are already skidding is effective but only to a limited degree.

Perhaps controversially, John Lyon holds the view, in the light of years of practice and experience, that the best way to slow down is to use the braking system. Most cars have front or rear-wheel drive, with two wheels applying engine retarding effort and control of speed, and only two steering; but brakes operate evenly on all four wheels. The difficulty is that any lack of delicacy by the driver under slippery conditions will cause a worse skid. However, delicate and controlled braking on all four wheels, *brushing* the brakes, will control speed and steady the car to an amazing degree. Experience gained from Scandinavian rally drivers has convinced us that the preferable technique is left foot braking, if you develop the technique on the skid pan first. Conventional right foot braking, used very delicately, is almost as effective. But there is the delay whilst moving your foot from the accelerator and you cannot balance power and braking

The benefit of ABS brakes is that they may save your life.

E514 ECF

together. Left foot braking, 'brushing' the brakes, is quicker and will enable you to control front and rear-wheel speeds. It assists the differential in the rear axle, if it has no limited slip, and blended with skilful application will control the difference in road wheel speeds. This technique is not for the inexperienced, nor for the public highway. Without constant practice your left foot will not be sufficiently sensitive, and you could cause a worse skid with excessive violent braking.

The table at the beginning of this section shows that wet surfaces usually have only 66% of the adhesion of dry surfaces; so judge separation distances and speed accordingly. Standing water, a 'puddle' to most of us, is very unpredictable and can set up powerful forces on your car. Wider tyres can aquaplane more easily and deeper water 'catching' one wheel with a strong retarding force, can spin a car. Again, never overreact, use the minimum necessary corrective action. If wheels do aquaplane or you skid, any sharp steering deflection that is maintained as your tyre achieves grip again will cause a sudden deflection of the car. Be aware of the angle of steering applied throughout any manoeuvre. Always try to 'see' standing water, changes in road surface, oil or other substances on the road that will affect adhesion, avoid them with an early deflection of course or reduce speed. Remember it is an offence to splash pedestrians.

The proper use of winter tyres and snow chains in certain weather conditions is important. Excellent as the best of the 'all weather' tyres are, they are not sufficient in conditions of constant snow and ice. In countries where you will be subjected to driving in such conditions, winter tyres or chains of the correct type are *essential*. Tests carried out by Porsche engineers at Weissach have shown the advantages of the correct tyres. A particularly striking example is the braking distance of a Porsche 944 with ABS from 50 km/h on an icy surface. Using one of the best winter tyres braking distance was 58 metres, but using spiked or studded tyres reduced this to 51 metres. On the winter tyres you would have hit the obstacle at 19 km/h. Although, unfortunately, the fitment of spiked or studded tyres is banned in many countries we recommend their use, where possible, if you are spending some time in the country in wintertime.

It is important to remember that winter tyres will not provide the same behaviour or adhesion on ordinary wet or dry surfaces. Winter tyres are specially designed for a purpose and should never be used on ordinary surfaces, unless it is unavoidable. The tread depth of winter tyres should never be less than 4 mm, below this a dramatic loss of tyre performance occurs. Winter tyres tend to lose their flexibility and hence performance after perhaps, two or three years. When replacing winter tyres always follow the manufacturer's instructions about 'mixing' tyres; if new tyres are to be mounted on only one axle, fit them to the front. However, rear tyres should *never* be run with less than 4 mm.

If you have to drive in conditions of heavy and persistent ice and snow, use the tyres or chains recommended for your particular make or model. Chains are a legal requirement in some countries at specified times. There are speed restrictions on the use of tyres or chains, and most manufacturers give technically permissible speeds; never exceed these limits.

The same care must be taken over the selection of chains as your tyres. Always ensure you use chains that are recommended by the manufacturer for your particular model, wheel and tyre size. Fit chains in good time, not when you have slid to a halt on heavy snow or ice. If your car has an ABS braking system, you will need special chains.

Finally, neither chains nor winter tyres in snow or ice, are a substitute for skill. Choose the correct safe speed and make all control movements with a delicate touch.

You Choose the Limit

Driving your car on the limit can only be done safely on a track or skid pan, under expert supervision. On public roads you must always have that margin of safety so necessary to safe, considerate motoring. Remember, others may not have such respect and will drive on, or even over, their limits; your margin of safety may save your life as well as theirs.

Regardless of the driving conditions there is no substitute for skill and a healthy pessimism as to the likely outcome of driving situations; anticipate danger, particularly in adverse conditions. On snow and ice, the situation is simply more critical and the correct use of speed and gentle, gradual, use of the controls is more essential. Aggressive braking, steering, or acceleration will produce only one result; a lack of control.

The proper use of the circuit and skid pan training has, in our view, other advantages, apart from increasing your level of skill and car control; it will most probably provide you with a healthy respect for your own limitations. Driving is not a 'natural' skill, although the male of the species, in particular, often regards any criticism as an exceptionally personal insult. However, pitch yourself against other Porsche drivers in the Driving Day handling tests and then you will not continue to have an overrated sense of your abilities. Watch how a master driver handles your car through the Porsche Customer Driving Day handling tests, or on the track, and you have a yardstick to measure yourself against. There is no substitute for expert tuition to develop your handling skills, your road sense, and tolerance and consideration for other road users. Knowledge of yourself and your own abilities is an essential prerequisite to learning more; we learn by first realising how much we do not know.

A complete understanding of the forces acting on your car and an appreciation of the limits of your handling abilities, gained under safe and controlled conditions, are an invaluable part of a driver's judgement. Too few drivers, in our view, have the opportunity to refine this judgement. It seems somewhat odd, that when driver error is a major factor in 90% of accidents, so little is done to reduce it!

You the driver are the one element that can 'choose' what speed, what action and what reaction to take. Your skill, consideration and judgement can make *any* situation safe, or dangerous. The choice is always with you, the driver; make it safely.

CHAPTER 7

EXPERT ROAD SENSE

The word 'sense' in this chapter title is as misleading as when it is used in the phrase 'common sense'. In neither case is it common, and it means much more than simply sense. 'Expert road sense' is a very rare commodity and only acquired by long experience, persistent application, sound training, and a deep desire for excellence. Expert road sense means, of course many things, and in the preceding chapters we have described many elements of it. In this chapter we are more concerned with reinforcing your ability to recognise hazards (actual and potential). To enable you to 'read' and 'blend' with patterns of traffic, to use speed safely and appropriately, so that your driving flows and you make safe, swift, and unobtrusive progress. We are not now concerned with specific techniques, but with your driving philosophy.

The physical action of driving is basically a set of very simple 'flowing' and coordinated actions, rather like flying. The complication comes when trying to perform either activity expertly, and we are *not* talking of race driving, but of the safe and expert handling of a high performance road car on the public highway. The very good driver will always drive safely, swiftly, and with a great deal of consideration for other road users, but the expert will achieve the very same result with greater precision; more economy of effort, of road, and with less stress on the car. Road sense is a matter of judging combinations of weather conditions, road types and surfaces, other road users, hazard potential and your own ability; always leaving yourself with something in reserve. You also need a deep understanding of the handling characteristics of how you and your motorcar behave, on the 'limit'. Then, with anticipation, planning, understanding and consideration for others, you have the basis of 'expert road sense'.

A driver with expert road sense will always make allowances for the actions and attitudes of other road users and will respond appropriately. You should never intimidate a bad driver, hang well back and give yourself 'room to work'. The pattern is easy to recognise; they slow down first, brake, signal (if at all), then look in the mirror, as if to see what effect their bad driving is having upon their fellow road users! Then, they fail to position properly and cause an obstruction – it's painful to watch. You will see it on the motorway too, as they fail to anticipate or judge speed and distance in the mirror. They pull out and signal too late, as they change lane, instead of four or more seconds before. They often drive very badly, without due care and attention, or reasonable consideration for other road users. An offence that, unfortunately may need either the supporting evidence of an accident or another witness, before action can be taken. It seems likely that our roads will continue to be populated by drivers with such attitudes, making any real progress in road safety difficult. Perhaps the proposed new offence of "very bad driving" will lead to improvements in road safety; who knows?

'Expert road sense' begins with complete mastery of yourself and the reasonable application of self-discipline. Then, when you know your limitations, the application of the techniques we have discussed becomes relevant. The consistent and conscientious application of the driving plan is important. Stages of the plan start with the mirror and all round observation. The *proper use of the mirror* is not simply a question of looking into it. You should always try to assess the *value* of what you see, then use your anticipation and judgement of *speed* and *distance* to determine your actions. Distance can be judged quite easily, but the accurate assessment of speed is much more difficult, as it involves time and distance. To help your judgement, look in the mirror at least twice and look ahead in between. The first sight establishes the distance involved, the second and any subsequent look will enable you to assess closing speed. Then, if you intend to alter course, signal to inform, to ask "Please may I?". Always give a fast approaching motorist *TIME TO REACT*; they are most vulnerable to your actions. Do not *instruct* others with the indicator, giving them no option but to react quickly and brake hard. Braking from high speed is a hazardous operation, particularly if there are closely following vehicles on an adverse camber, or curving road. Then you should consider delaying the signal without closing your own safe following distance.

ON MOTORWAYS OR DUAL CARRIAGEWAYS, any signal to turn or change lane should be given very early without a change of course or speed, again asking "May I?". You must always give time for others to understand your signal and to react to it. Your signals should operate for at least four seconds before you decide to alter course; and six or seven seconds at high speed. On motorways, extend your planning horizon and use long sight lines; they should be more than sufficient for you to spread out your following distance, signalling time and steering deflection – it's a big place so use it. You *must* see evidence of reaction from the driver behind before you can change lane with safety and convenience. If you assess that you can safely signal your intention to change lane to a closing vehicle, do so positively and without delay – don't be timid and don't delay, ask! Accurate judgement of speed and distance ahead and behind is vital, so don't expect strangers to read your mind – communicate. If necessary, wait for a safer time, but remember that waiting to signal may cause more danger, as you arrive later into a hazard. Try not to give other drivers the 'chop' by late signalling due to poor anticipation or, dare we say it, intimidation? After overtaking on a motorway, you should not normally be required to give a signal at speed as you move back into the centre or nearside lane, as your lane change should never be close enough to disturb the driver of the overtaken vehicle. It may be necessary to signal in slow traffic queues. Remember, nearside passing must not create headway, or it is *overtaking* and is a traffic offence.

Following expert drivers is a rare and welcome privilege; it should be educational too! Their signals are always early allowing others time to react, before they change lane or course – a technique which is always appreciated. Signal early yourself and see evidence of reaction in the mirror, only *then* can it be safe and convenient to change lane. Cancel your signal before you cross the lane line; it is given as an *intention*, so there is no value in continuing the indication once your manoeuvre is completed. Don't forget a 'thank you' wave afterwards, if you have received cooperation. Pulling out and signalling at the same time is often self-righteous and bloody minded; you are giving an instuction. If you make a mistake and miss seeing a driver in your door mirror, there is no time for him to react, so always give a life-saving shoulder check, after an early signal.

Most modern cars have excellent all round visibility, particularly those with interior rear, and external near/offside mirrors. However, mirrors can lie so always use the shoulder check! Concave mirrors will make following objects appear *larger* and therefore closer than they are, whilst convex mirrors will make them appear *smaller* and therefore further away. Use of mirrors should always be before giving a signal, checking that there is a reason to give one. If you always check that there is a reason for a signal you will *know* whether there are other vehicles, pedestrians or cyclists in the road scene around you. However, if there is any doubt of your ability to see them it is better to be safe than sorry, establish the reason in your mind, make proper use of the mirrors and signal if you are not sure. This all round observation and awareness should be conditioned by acute reasoning to signal, very important for safe driving. Even when setting off, always check the blind spot, for an entrance out of sight of your door mirror, and signal before emerging from behind stationary parked cars. Always make a final look before you set off as it takes time to manoeuvre, and you are vulnerable when so doing.

Always communicate with other road users in the road scene who can see it, timing your signal for their individual situation; it is always changing. Even at the same junction, the next day the situation could be totally different and your signal will require new timing. Don't flick the indicator automatically, but *think* before you act, and then

(Above and opposite) Positioning, patience and self-discipline are vital ingredients to expert driving. Signals should be positive and early.

act with expert judgement and control upon what you see. It is unnecessary, in most cases, to signal when passing a parked car and it can be confusing and even dangerous for following vehicles if you are approaching a right turn. In the same situation, but with a fast motorcycle approaching behind, an early, "I intend to move out" for a parked vehicle way ahead, asking "Please may I?", and seeing evidence of reaction before steering deflection, will be invaluable. It is then cancelled before passing and long before any future right turn, so as not to confuse.

The left turn indicator is often used to say, "I intend stopping on the left", so do be careful about signalling for this reason before a junction, since a driver may emerge from the junction assuming that you are turning left. Tapping the brake light on and off to signal "I am slowing or stopping" may be used to give a sense of urgency when slowing from higher speed; a sensible precaution. Drivers who use hazard lights when parking illegally or unloading and those who use high intensity rear fog lights when visibility is *more* than one hundred metres, are using them illegally.

Rare occurrence though it is, a smartly given, clear arm signal can be invaluable at times and never more so than when approaching a pedestrian crossing in town. This is where sound judgement of the 'all round' situation in the mirror and the hazard ahead is imperative. In a busy high street an arm signal is more deliberate and incisive than an external brake light, although it is occasionally better

to use both, which requires expert anticipation and judgement of speed and distance. The arm signal, "I intend to slow down", given neatly and correctly, is only two sweeps of the whole arm from the shoulder height to the door panel, palm to the ground, wrist straight, returning both hands to the steering wheel *before* the stop lights come on. The continental arm signal, "I am stopping", arm vertical from the window ledge, palm to the front – is far more realistic. An arm signal to turn right, is given with the palm facing the front, fingers together, to reflect light (not a pointed finger), for at least three seconds. Don't give arm signals unnecessarily, or at speeds above fifty miles per hour. Perhaps you should signal your turn into a concealed entrance by arm, very deliberately and early, then by speed (slow early) and position (to the left or right of your lane). Be very positive and deliberate to avoid any confusion to following traffic.

The turn left arm signal should be executed with the right arm, as follows. Keep the wrist straight, palm to the front, and circle around from the elbow, resting the arm on the window, and give at least three circles of the arm. Another awkward arm signal displayed in the code is "I want to turn left" to persons controlling traffic – left arm straight out across your passenger's face! Continue to give the old one if necessary, with the right arm across the chest, with the back of the hand giving sufficient clarity, but without rudeness to your passenger. If required give this signal on approach to a driver emerging from a left turn, to be more deliberate.

Never give instructions to other road users, or reverse priority at junctions, no matter how helpful you think it may be. Just stop gently, or leave the gap for others to turn in a stationary traffic stream, and keep a blank expression on your face! You could be liable if an accident occurs – you will not see a policeman sitting at the wheel of a car give instructions to traffic.

It is difficult to understand the driver who accelerates to beat a green light or accelerates over a stop line with amber showing and the mirror clear. If a green light has been showing for as long as you have seen it, it is reasonable to anticipate it will change and you should prepare to stop by easing off the power. This not only increases passenger comfort, security and well being; it is much safer! Remember, if the green light fails you must obey the secondary signal as if it were the primary, but once clear of the stop line, you can continue with care. This problem causes much confusion with the inexperienced, as does the yellow box junction when turning right. You may enter the box to wait to turn right if your exit is already free of stationary traffic sufficient to accommodate the length of your vehicle and be clear of the junction.

At a railway level crossing, even if the red lights flash for more than three minutes without a train approaching or the half barrier is down without the lights, don't cross! If your car breaks down or stalls only halfway, get yourself and your passengers out of the vehicle *quickly* and phone the signalman. Then, only if there is time, push the car clear keeping a sharp lookout – don't be stubborn or take chances.

There are three forms of visual signalling you can use; hand, indicators or lights. *The latter* are excellent if used with a sense of timing, deliberation and discretion. Use them when approaching 'blind' junctions at night or before overtaking, to warn of your approach. Headlights are often abused, used wrongly and aggressively, causing very 'obtrusive' progress. Never 'flash' your lights, using them to call other road users across your path, or to reverse priority. Like any other signal, they should be given clearly, decisively and for as long as it is reasonable for them to be seen and understood. Normally, the headlights will need to be used for at least four seconds, before there is positive reaction. Only use your headlights in daylight if absolutely necessary, or when the horn will not be so effective. For example, approaching minor junctions at a distance and where there is a road user about to pull out. Similarly, if you are approaching other cars and you give an indication (or you feel a warning necessary) of overtaking when there is a long distance prior to the manoeuvre; use your lights not your horn. However, don't annoy people by using the warning unnecessarily.

Use the lights in 'dead' sound conditions when it is raining or to penetrate the peripheral vision of the lorry driver in a noisy lorry or tractor cab. He or she may be wearing ear defenders, so look for evidence of reaction before committing yourself to overtaking – they may not see or hear the warning, and watch for that slight movement of the offside wheels away from the centre line, as evidence of reaction. Don't flash the headlights several times aggressively when someone is already overtaking and obviously can't move in – that is just bad manners and will antagonise many people. Give them fair time to move over, even if they are being obstructive. Then, if there is evidence that they haven't seen you, use a four-second light on main beam to attract their attention.

Don't close up and intimidate a preceding vehicle, forcing them to move over. That's reckless and potentially dangerous, a similar offence as overtaking on the left. Try to be sure you don't use the headlights, excessively, in the face of oncoming traffic in difficult light conditions. But, if you have good reason to use the headlights do so, don't always expect others to be aware of your presence or to be cooperative. Communicating with a clear headlight warning, like any other signal, should be given early, clearly and decisively; it is better to give a warning than have an accident.

On single carriageways when overtaking more than one vehicle in a queue, use the headlights from well back. But don't commit yourself from too far back and 'charge', or be aggressive. 'Pick' your way one vehicle at a time, if necessary. Headlights may be better than the horn in this instance – to be seen by all. The turn indicator is useless to attract the attention of others in front of you, only use it when you intend turning or altering course, if necessary. Remember headlights have the same meaning as sounding the horn, they are a visual warning, an indication of your presence. They are not to be used to give an instruction.

Do not forget the visual signal of an acknowledging wave with your left hand (it can then be seen in the middle of the screen of a right-hand drive car). Use it only in acknowledgement of the courtesy and cooperation extended to you, it will do much to foster a spirit of road safety in others. A short, positive wave is all that is required. At speed, keep both hands steering and slide the left hand up towards the top of the wheel and raise the fingers and palm from the rim. After overtaking someone who cooperates, you should acknowledge their cooperation when you are completely safe and back on your own side of the road. Courtesy promotes good road manners and engenders that spirit of chivalry so badly needed on our roads today. The most extraordinarily selfish behaviour and attitudes of mind are developed by road users that have nothing to do with the spirit of The Highway Code, are very dangerous, and can make even driving a Porsche a misery.

The most prevalent international disease is TAILGATING; following far too closely to be able to stop and avoid hitting the car ahead if it does a 'tyre smoking' stop – particularly in a 'crocodile' queue that concertinas through delayed driver reaction. After all, nothing

Tailgating. There are no good reasons for this, only bad ones that kill.

stops as quickly as a small car that hits a stationary vehicle! *There are no good reasons for driving too close to another vehicle, only bad ones that kill!* Tailgating by the very bad driver is an attempt to intimidate the preceding driver to speed up or move over, or wilfully to prevent a driver from filling the gap. In our view, such action positively promotes an accident situation, and often an equally aggressive response. Much better to hang back and use the headlights to attract the attention of the unaware for at least four seconds, but only when the driver in front has a good, safe and convenient opportunity to retire. Let's face it, if a driver has neither the opportunity nor desire to overtake, a safe following distance should be there to accommodate the person who does! We have mentioned this problem before (and probably will do so again), but aggressive intimidation by following too closely causes fatal accidents.

Use of the horn to give an audible warning is, in the correct conditions, the best means of indicating your presence. Why is it then that in this country many drivers object to sounding the horn? In correspondence in the I.A.M. magazine, several drivers actually boasted that they have never used the horn! Is it a natural British instinct to be discreet, to try and avoid drawing attention? It often seems that some would rather have an accident than make a noise! No matter how slowly you drive, other road users may not see or hear your approach and not take the necessary precautions. Some use the horn to retaliate when some offence or inconsiderate driving has been committed – often the only occasion they do use it. Perhaps most abuse of the horn comes from the driver who drives too fast in towns for others to react and uses the horn as a substitute for slowing the car. Then the horn is used aggressively, demanding 'right of way' from anyone who dares not to cooperate. Used in this way, the horn hardly encourages a spirit of good fellowship on the road, and certainly reduces its proper contribution to road safety. The following question is invariably included in a motor insurance claim form – "Did you sound the horn?" Perhaps they should add, "If so, when, and was there any evidence of reaction to the warning?"

In heavy town traffic, occasions for using the horn are rare – speed should be low, so that other precautions can be taken quickly. In lighter traffic, the need to give a warning will increase and a light 'tap' is often enough to attract the attention of the vulnerable, unwary pedestrian or cyclist, but only if approaching traffic drowns the sound of your approach, or if they show signs of carelessness.

As speeds rise in the country, the need to give a warning will be seen more as a necessity on approach to completely blind bends, crests, junctions or entrances. If you cannot be seen, why not be heard? But good judgement and discretion are still required. Do not sound it for every unseen area of danger – it depends upon their importance, and whether an alternative driving plan can be adopted. Inexperienced drivers often have difficulty in noticing and assessing actual hazards within vision and will rarely see the need to sound the horn for concealed hazards that can reasonably be expected; pedestrians, horses or cyclists, around narrow 'blind' bends for example. Such anticipation develops with experience.

If there is no evidence in their behaviour that a road user is aware of your presence, then warn them if necessary. *Communicate – use the voice of the car*! Sound the horn once, timing its use so that it can be identified as being for the person intended, and for them to have time to react. Sounding the horn several times will give an impression of aggression, except for children when several light taps may be necessary.

There are three types of audible warning: the tap, a medium warning and a long continuous warning note when approaching at high speed (for the note to carry). Be sure of a safe reaction before you commit yourself to the hazard – a road user may be deaf! In the country pedestrians, equestrians and cyclists will hear you coming with no approaching traffic – you must refrain from making an unnecessary noise. Just steer a wide course if there is room, if not, *wait* until there is.

When motoring quickly, you will close up with other road users rather fast. Try not to surprise them by overtaking at speed without considering whether they are aware they are about to be overtaken. A sudden rush is likely to disturb them, particularly if overtaking is

from behind or from too far back. Always consider the time they need to react and cooperate with you.

Warning of approach is the last feature of a trained motorist's approach plan to a hazard – only to be sounded after every other safety precaution has been taken. With the car steering on the correct course, travelling at a safe speed and in the correct gear; *then* is the time to consider sounding the horn.

The keys to expert motoring are many but one of the most important is the judgement of appropriate pace throughout your journey. Using the driving plan carefully and consistently, you will begin to flow with the traffic, by developing your sense of speed and acceleration. Never sacrifice smoothness for speed. Use the accelerator to bring you to situations as they 'open up', holding back when it is safer and appropriate to do so – a sign of true mastery of yourself, your car and the road conditions. Acute and continuous observation, coupled with this accurate use of power will enable you to 'flow' through traffic, reducing the need to brake. When a hazard that requires braking appears, recognise it early and *spread out* your braking.

Every road has a pattern of hazards that form its general character – if you appreciate this, you have road sense. The A41 has a similar character through its length, so has the A4 or the A5. In all road patterns, try to constantly vary speed with power for changing conditions and road surfaces. But you need to apply your 'road sense' before

(Opposite and above)
When closing up on other motorists, always be sure you have been seen.

you get into the car, particularly for journeys over strange country or on strange routes. Route planning is where expert road sense begins. Plan your route for specific objectives whether they are scenic, speed or simply interesting reconnaissance for future driving. With practice and experience you will be able to 'sense' routes that will satisfy your objectives. Always try to take account of the wishes of your passenger(s); better that they are calm and content rather than distracting you. No one knows how often the distractions of passengers have caused accidents. Those of you who have had to drive long distances with young children or very talkative adults, will know what we mean. Much better to give everyone regular leisure breaks for their comfort and safety.

Speed should always be governed, first and last, by safety; better to travel safely *and* arrive. On long cross-country journeys a fifty-mile per hour average is quite high. Remember you do not need many traffic holdups in towns or villages to slow you down. On motorways averages will, of course, be higher. Nevertheless, learn to pace your journey and always give yourself time; plus an adequate contingency. Include stopping times in your estimates for average speed; remember if you stop for 10 minutes in an hour you will be 10 miles behind a 60 mph average and you would have to average 80 mph for 30 minutes to bring your average over one and a half hours back up to 60 mph. Far better to plan your drive properly and with adequate stopping periods; accidents can take up a lot of time, and some last for ever!

There are more than three hundred thousand accidents on the road every year in this country, costing some £400 million, and a quarter of them are directly attributable to skidding. The knowledge and skill gained on the skid pan, with proper skidding instruction, would be invaluable. Unprovoked skidding on a slippery road is normally the result of bad driving, failing to anticipate the conditions, and often reacting in a manner which increases the problem. There is no substitute for actual experience, in an absolutely safe environment. You will never regret making use of the opportunity to drive on a track or skid pan, to develop an understanding of the car "on the limit". Not that you will *ever* drive to such limits on the public highway. This knowledge gives you the reserve of skill in the event of an emergency. You are then able to react with the minimum necessary applications of brakes, power or steering and are more likely to retain control. This confidence of your ability to control your car under all conditions can save lives, if used properly in an emergency. It should never be used as an excuse to drive too fast or near the limit. Judgement and an almost pessimistic sense of risk and responsibility are attributes of all expert motorists.

The expert driver *always* matches speed with vision; is always able to stop safely, within his or her line of vision. Perhaps this single rule is the most important element of safe motoring; certainly many

Read the road and always match speed with available vision.

accidents are caused by drivers not obeying it. Unfortunately, hazards are not always visible, but the expert driver will be able to recognise situations where hazards can reasonably be expected, and will react accordingly. This, almost extra-sensory power of anticipation is really no more than acute observation, matched with learned experience. The expert driver uses his or her knowledge of what is proper to anticipate hazards.

Care of your car is an obvious and vital aspect of your motoring. Not simply to apply the standard checks but always to know that your car is in perfect mechanical condition. Regular servicing by Porsche trained technicians will maintain your car's value and its performance. Never apply minimum standards of maintenance, always follow the manufacturer's instructions for service intervals and replacement parts. Your tyres should always be regularly checked, and maintained at the correct pressure. We recommend that you replace tyres as soon as wear reduces the tread to 3 mm.

Planning, deliberation, skill and 'flair', are all necessary attributes of an expert driver. Planning and discipline is the basis of all expert driving and is needed to apply the skills and techniques we have discussed in the preceding chapters. But experience and skilful tuition are also critical, particularly structured and learned experience. Any fool can drive 50,000 miles per year and learn nothing! To develop your driving requires enthusiasm and a real desire for excellence. Perhaps, somewhere, there is the 'perfect driver', but we have never yet met him or her. Although we cannot prove that he or she does not exist, we doubt it!

This book is not a replacement for '*Roadcraft*' which we believe should be mandatory reading for *all* drivers. Our objective is to help you develop your skills and share some of our enthusiasm for driving. Personally, we are convinced that the Driving Test should, clearly and unambiguously, be the first stage in everyone's driving. Remember "Life is not a rehearsal, nor is it for beginners!" Maybe we should all consider ourselves as 'learner drivers', and remember to go on learning. Unfortunately, it is the worst who think "they know it all". If only they would remember that road accidents are not rehearsals either; for some 5,000 people a year they are the 'Final Curtain'. Although drugs, alcohol and mechanical failure contribute to 30% of fatal accidents, that leaves 70% down to human error; people thinking they are rehearsing? In the Police Driving Manual '*Roadcraft*', there are Ten Commandments of Motoring, we quote them for you to remember:

1. Know the Highway Code and put it into practice.
2. Concentrate all the time and you will avoid accidents.
3. Think before acting.
4. Exercise restraint and 'hang back' when necessary.
5. Drive with deliberation and overtake as quickly as possible.
6. Use speed intelligently and drive fast only in the right places.
7. Develop your car sense and reduce wear and tear to a minimum.
8. Use your horn thoughtfully; give proper signals.
9. Be sure your car is roadworthy and know its capabilities.
10. Perfect your roadcraft and acknowledge courtesies extended to you by other road users.

The qualities of the good driver are certainly not the subject of unanimous agreement, but the above rules will certainly be followed by the expert driver. The only factors we would add are the *desire* to be an excellent driver, and have flair. With your Porsche you certainly possess the automotive technology; do you have the personal skill and consideration to use it? Expert road sense, can be demonstrated, it can be tested, but it can only be acquired if you the driver, wish to apply it consistently. The desire for excellence is an absolute necessity, as is a healthy respect for your own limitations.

CHAPTER 8

DRIVING THE 'CLASSIC' PORSCHE: THE PORSCHE 911

HANDLING THE EARLY LEGEND

The publication of this book, and the 25th anniversary of the Porsche 911 coincide. An appropriate coincidence, since THE PORSCHE, that in a quarter of a century of continuous development has remained a classic driver's high performance road car, is truly "Driving in its purest form". The Porsche 911 has retained both the original body shape conceived by 'Butzi' Porsche in 1956, and the rear engine, six-cylinder 'boxer' layout. Twenty five years' success on road and track has sustained this car's position amongst the world's true 'supercars' and, to some of us, it continues to offer the ultimate in high performance motoring.

A major attraction is the car's straightforward practicality, the fact that you can shop with it, and then drive off for a long, swift, continental journey. Perhaps even enter a race or rally? Most of us would not want to go that far, or fast, but this car has proved its ability to do both; all with a twelve thousand mile service interval! The handling and performance of this modern motoring legend (a legend in its own lifetime?) has remained ahead of its years. The first Porsche 911 Coupé had a 2-litre engine, disc brakes on all four wheels, and delivered 130 bhp at 6200 rpm. The current model Porsche 911 Carrera Coupé, has a 3.2-litre engine which delivers 231 bhp at 5900 rpm. That's Porsche development!

The handling characteristics of the early 911s were certainly not as 'sensitive' as sometimes imagined. As Paul Frère wrote, after testing an early factory demonstrator across France and Italy in 1966;"on the road cornering speeds were generally limited by visibility and traffic, rather than by the actual cornering power of the car".

The truth is that the car's handling limitations were far higher than the average sporting motorist required. This was, and is, due to the fact that the 911 owes much to Porsche's early racing experience with the Auto Union P-Wagen and the rather hybrid 356, and this long racing heritage contributes to the car's tremendous 'feel'. Furthermore, the design was the product of intense preparation and development, and pre-production investment reached some DM15 million; an enormous sum in the 1950s. Unceasing development, building on the car's superb basic design, has given reserves of handling and power, that are now greater than ever. This emphasis on 'active safety', reserves of handling, grip, braking and acceleration, gives very large safety margins; even to the inexperienced.

The handling improvements made in the first four years of the production of the Porsche 911 were truly outstanding. The extent of the achievement of the devoted team of Porsche development engineers and test drivers is, perhaps, best illustrated by the fact that the 1968 'B' Series cars achieved lateral acceleration figures of 0.897g, on the 190-metre circular test track at Weissach. These figures should be put into the context of performance figures of the period when 0.65g was considered good for a road car, and 0.7g excellent for a sports car; Porsche was certainly way ahead of the field. These figures should place the early roadholding of this magnificent car into a more realistic perspective of cornering performance than you would believe from other accounts.

Interestingly, Porsche engineers at Weissach consider that some 80% of subsequent development has been due solely to improvements in tyre technology. The current model's outstanding handling performance, in comparison with later designs that also benefit from this universal advantage, proves the brilliance of the original concept.

At Weissach we discussed the matter of the handling characteristics of the 911 with one of the engineer test drivers. He emphasised that, in the 1950s and 1960s, the desirable handling characteristics of a current racing or sports car were very different from today. Because of tyre, suspension, braking and engine technology, oversteering characteristics were required for fast driving. This is quite evident if you watch films of professional and amateur drivers of the period, whether racing or rallying, on road or track; the four-wheel drift was the technique, and it looked spectacular. Today, the desirable characteristics for a modern high performance road car are for a basically neutral to an understeering car, with a smooth and predictable transition, at extreme cornering forces, to oversteer. Watch a first class modern saloon car driver or a Grand Prix driver; smooth, clean and very tidy. All that excessive oversteer does to a modern sports racing or Grand Prix car is to scrub off speed, and tyre rubber; poor technique.

The company's pursuit of improvements in this, the first and most important, phase of the car's development is a typical example of Porsche's consistent and meticulous policy of product development; the best is never good enough, perfection is desired. Since 1963, the 911 development programme has been coupled to a very active racing programme, both for factory and privately entered cars. This continuous involvement in racing to improve the breed is, in our view, unique amongst the world's producers of high performance road cars. Evidence of the early success of this programme is that two Porsche engineers, Falk and Linge, took a 911 to a class win in the 1965

The famous 'Duck tail' Porsche 911 Carrera RS with the 2.7-litre engine.

The inner workings. A cutaway of the Porsche 911 Carrera with Sport Equipment.

Monte Carlo Rally, on ice and snow. Then, in 1968 a Porsche 911 T won the event outright, driven by Vic Elford/David Stone, whilst another 911 T came second driven by Pauli Toivonen, who went on to win the European Rally Championship of that year.

This remarkable high performance road car has, at various times, suffered from premature obituaries. By a process of continuous development Porsche has ensured that this car's handling, safety, performance, and technology, is second to none. The car is a magnificent example of a good basic design that remains aesthetically 'right' with continuous technical development ensuring that it always lives up to its design. Launched at the Frankfurt Motor Show in September, 1963, this car's 25-year development programme has ensured its place as one of the world's great high performance road cars; for some 'the greatest'.

Handling The Current Porsche 911

This whole book is concerned with excellent road driving skills, and not, repeat not, with driving on the 'limits' of any car or driver, except to illustrate appropriate handling skills and gain necessary experience, in a completely safe and appropriate place. Therefore, before discussing the handling qualities of this exceptional motorcar in general, it is perhaps worth emphasising a number of points. Firstly, as we have said in a previous chapter, the basic handling characteristics of a car are a function of the layout of its major components. However, with modern tyres, suspension, and braking technology, engineers can, and do, alter those basic characteristics to an extraordinary degree. Furthermore, if you allow the tyres, suspension system (especially shock absorbers), or brakes, in particular, to become worn or suffer from a lack of maintenance, you will alter the car's handling characteristics, with potentially dangerous consequences. In our opinion, if the 911 is driven in the expert manner we have detailed in the previous chapters, matching speed with vision at all times, assessing the appropriate speed for corners, using the controls smoothly, in the manner described, in normal driving conditions, you will never utilise the enormous reserves of braking and cornering power of this car.

The current models are exceptionally stable and display all of the very best characteristics of a modern high performance road car. The tremendous traction offered by the rear engine, rear-wheel drive design means that all the power can be used in safe circumstances. This not only provides great reserves of acceleration in the right place but also gives grip and traction in adverse weather conditions. The suspension, brakes and aerodynamics are all more than capable of dealing with the exceptional power and torque of the well proven, and wonderfully smooth 'flat-six' engine. The servo assisted braking system has been designed to match the 42% / 58% front to rear weight distribution (38% / 62% on the 911 Turbo), with smaller servo pistons on the front brakes and pressure limiting valves in the system reducing any tendency to lock-up. The car's steering is not servo assisted and is suitably sensitive and direct, with the correct balance and 'feel'. The aerodynamics, at higher road speeds and particularly in severe cross winds, give a very high degree of directional stability and the suspension will give a very flat, stable ride, even under extreme cornering. It is a complete high performance road car that, depending on your preference for a particular model, will satisfy the demands of the most expert driver.

Let us consider how this car's handling changes, progressively, on a constant radius circle, similar to that at Weissach. We will, of course, assume that all of the comments about tyres and maintenance are taken as read. The 911's initial tendency, on any flat corner, at low speeds, is to display neutral characteristics; it will go exactly in the direction that you point it. Increase speed, with a smooth and progressive application of a small amount of throttle, and maintain the desired speed with a constant throttle, always pushing the weight of the car. You will notice that understeer will start to increase as the rotational speed, and lateral acceleration, build up; you will have to increase the deflection of the steering to maintain your position.

At Brands Hatch, the Porsche 911 Carrera with Sport Equipment shows its understeering characteristics.

As you inexorably increase speed, at a very slow rate, the understeer will build up, more and more. Continue to keep the engine pushing the weight of the car, do not lift off the throttle, and understeer will simply go on increasing. At a very extreme angle of understeer, when you are steering into the corner at a very high slip angle, you may decide that that was enough. Instead of sustaining the car's speed round the circle you ease off the throttle, to reduce your lateral acceleration by slowing the car down. But you have an extreme angle of understeer on the car; the front wheels are pointing into the circle at quite a sharp angle. As you reduce the 'pushing' power of the engine, the weight of the car is transferred towards the front, onto the front wheels. The understeer will stop, both because of the weight transfer and the fact that the driving force (which contributes towards the slip angle) has decreased.

Remember the speed difference between these two conditions, tracking round with a constant throttle and radius at quite a high speed and cornering force, and then suddenly lifting off the power and turning inside the circle on a reducing radius, is probably only six or seven miles per hour. *To maintain the radius of your circle of rotation, you must immediately take off steering lock and point the front wheels along the circular path you wish to sustain.*

If you do not, the weight transferred to the front wheels and the reduction of speed increases their grip and the car will immediately and rapidly turn towards the centre of the circle. If you are cornering at sufficiently high 'g' forces and fail to remove the understeer, then the 'unweighted' rear will tend to come round. Realising what is happening you will, of course, steer into the skid; everyone remembers that. But it is too late – you should have recognised that this very responsive car was already recovering, in response to your action in removing the source of the speed, and you should have corrected the understeer immediately. Naturally, overreaction at this stage to the resulting oversteer will simply make matters worse. The important thing is not to be late in correcting the understeer which must also be done instantly you ease off the power or the tail will swing. If you were to continue to increase the power, you would find that the understeer would increase. It is not the oversteer that is the problem but the failure to react quickly, and point the steering on the path you wish to follow the very instant you remove the speed.

If you have a car fitted with the optional extras of stiffer 'Sport' shock absorbers, wider wheels, and ultra low profile tyres, there will be some differences in the handling. You will find the transition from understeer to oversteer will occur at higher slip angles, be considerably more difficult to induce (wet or dry), and be that much more predictable and progressive.

The steering on the standard 911 does not have any servo assistance, nor does it need it. The car's steering gives exceptional feel under all conditions, and the road and steering angle can be felt all the time. This is another advantage of the rear engine, rear-wheel drive layout, as with less weight over the front wheels the steering torque is very reasonable, even when parking. There is quite a lot of castor on the steering, which means there is a strong tendency for the steering to self centre. This also gives tremendous 'feel', but it does mean that the steering wheel tends to kick on rough roads, and as you go over potholes. Naturally, there is a cost for all this 'feel', which is to keep both hands on the steering wheel when cornering or braking, as we recommend for all cars, particularly when on rough or slippery surfaces.

Even if you have not seen that the road conditions are either wet or slippery, due to ice or say, oil, the car's steering will rapidly tell you. In these conditions the characteristics we have described will be emphasised, and the car will need to be driven with skill and sensitivity; the excellence of the car demands it. Porsche has built tremendous feel into the car, so that it tells you all the signs of too much speed, steering lock, or a suddenly changing road surface. If you experience standing water, which will drag the steering to the right or the left depending on where the water is, it is important not to overreact. If you have overreacted, and applied too much steering lock as you go through the water, as soon as you come out of it the superb grip and traction of the front tyres will bite, and take the car to the right or the left; just stay calm, the car is perfectly capable.

However, in conditions of extreme and heavy rain the car's weight distribution gives a tendency to aquaplane at very high speeds. The rear weight bias, particularly under hard acceleration, will provide exceptional traction at the rear wheels, but reduced traction at the front. This can lead to extreme understeer and, on occasion, the tyres will aquaplane particularly if you have very wide ultra-low profile tyres. However, this will only occur when there is a great deal of standing water or torrential rain, and you use the car's immense power thoughtlessly and recklessly.

The car is, as we have described, a fundamentally understeering car with tremendous balance and feel. The brakes are capable of clawing the car down from speed, extremely quickly and with absolutely no signs of fade or imbalance. The car is so well balanced that it is very possible to trail brake through corners, if you leave your braking too late, which of course you should not. The point is that the comment that you cannot brake in a corner, is simply not true. The car is so well balanced that, providing you maintain control of the steering, and are aware of all that we have just described, gentle and controlled braking will not unbalance the car.

The different 911 models will all handle in the manner we have described, but there are differences, depending primarily upon the wheel/tyre and suspension specification. There are two major differences that you will notice, depending upon the chassis type that you choose. The standard Cabriolet is, of course, rather more lively and tends to feel the road rather more than the standard Coupé. This 'feel' of liveliness is less pronounced in the 911 Cabriolet with Sport Equipment, because of the stiffer suspension and wheel/tyre specification . They will all have the same basic understeering characteristics but the models fitted with Sport Equipment will tend to understeer at marginally higher cornering forces. The transition from understeer to oversteer will take place slightly later and will be more difficult to induce.

Perhaps a final comment about the limits of this Porsche, a comment which applies to all models of the 911 Series, will emphasise the extraordinary abilities of this superb driving machine. In our view, anyone who experiences anything other than gentle understeer when driving the car on the road, will be travelling so far above any reasonable speed limit, and failing to match speed with vision, that they would probably be driving recklessly. This is only because of the car's extremely high handling limits.

Driving The 911

The question is which 911? Personal choices are always biased, so we will test drive the Porsche 911 Carrera Cabriolet, examine its characteristics, and give ourselves considerable pleasure in the process. Following the advice in the chapter on planning your driving, we have planned our journey carefully. We will know where we're going, and have a 'picture' of the journey in our mind. The weather

Porsche 911 Carrera cockpit.

is fine and dry. The car is a standard Porsche 911 Carrera Cabriolet, with a 3.2-litre engine producing 231 bhp at 5900 rpm. Approach the car from behind, then you can check that the rear lights are intact, and that there are no hidden obstacles, especially young children. Fluid levels have all been checked, fuel topped up, tyres checked and windows, interior and exterior, are crystal clear.

Remember that the pedals of the Carrera are slightly offset to the left; be aware of it and you'll soon find it does not affect you at all. The starting procedure is easy. Simply follow your 'cockpit' check (see Chapter 1) and turn on the ignition; immediately the wonderful flat-six 'boxer' engine sings into life. The boxer engine, so called because the pistons 'push' against each other in the horizontal plane, has a unique sound and even at tickover is unmistakable. Check the oil contents gauge, left-hand dial, right-hand gauge; it should read near to the top on level ground at idling speed and operating temperature; never take a reading when the car is in motion – the pointer will drop as engine speed increases. Consult your handbook for correct checking procedures. The oil pressure on the right-hand of the instrument to the left of the tachometer should read at least 3.5 bar at 5000rpm, preferably above. The classic lines of the panel are very welcome, compared with some of the more 'modern' vehicles. The tachometer is right in front of you, above the steering wheel hub, and the engine should idle just above 800 rpm. Check that the warning lights for battery charging and oil pressure are out; they are located between the oil pressure and oil temperature gauges. Let the oil temperature rise enough to register, and we are nearly ready.

Doors shut, seat and mirrors adjusted, seat belts fastened, comfortable (?), then let's go. Hand brake off, and as we're on level ground a static brake check. Clutch down, into first gear, 'feeling' the gear in gently as everything is still cold, then let the clutch bite and gently move forward.

Once on the move, and the engine fully warmed up, the true character of the car is revealed. Firstly, it is a beautifully balanced, progressively understeering car. Whilst it has exceptional straight line stability, the chassis is very much 'alive' and sensitive, giving you a complete picture of the road, without being nervous. The car's steering is exceptionally precise and there is no excuse for not placing the car exactly where you want it to be.

Paul Frère's comment on the early 911 that "road cornering speeds were generally limited by visibility and traffic, rather than by the actual cornering power of the car," is no longer true; road cornering speeds are now totally limited by visibility and traffic!

There are, however, a number of factors which the car's handling characteristics should encourage you to concentrate upon. The car demands excellence from its driver when driven quickly; all of the qualities of the excellent driver are a necessary condition for swift, safe progress. Anticipation is essential, particularly in setting the car up for hazards. Early positioning, and judgement of speed for corners will ensure that you can sustain power and therefore drive through the corners to sustain the basic understeering characteristics of the car. Sympathetic handling, or finesse, is also necessary, in all aspects of controlling the car's steering, braking or acceleration.

This 'finesse' is particularly important when cornering swiftly – it does require the driver to develop his or her ability to anticipate the course, position and speed required for a hazard. Cornering must be executed in a smooth 'flow' of power and steering, with sufficient power applied to maintain the speed, and sustain the weight and power on the rear wheels; this will maintain the car in an understeering condition in all but the most extreme cornering situations.

In the wet, the car handles in exactly the manner we have described but the limits are lower; remember the table of adhesion values? The limits of the car's exceptional cornering ability are unlikely ever to be reached in normal road driving. However in the wet, or on ice or snow, the understeer will be more pronounced and loss of adhesion is more likely but again only in extreme, driver induced, situations. In these extreme conditions the limited slip differential can be a very great advantage. Porsche engineers also recommend the use of proper winter tyres in conditions of continuous snow and ice, and there is little doubt that the limited slip differential option offers substantial advantages of traction and grip, which will be enhanced when used in combination with such tyres.

Opposite Porsche 911 Carrera Cabriolet. "Pure driving pleasure".

A 911

Where you have to be extremely watchful is when you approach a corner hurriedly, far too fast for conditions of limited grip, failing to match speed with vision. Then, the instinctive, natural, and often abrupt 'lift off', from the accelerator will induce the loss of understeer we described earlier. The loss of drive will throw the weight of the car forward onto the front wheels, reducing understeer, and 'unweighting' the rear wheels. If the cornering forces are extremely high, or the road is wet, you may well lose adhesion of the rear tyres and, in a cornering situation, the rear end could break away. So take the power off gently, power is the cause of too much speed – remove the cause, but not too harshly. Then if necessary, instantly, correct the steering and point in the direction you wish to travel. This will only be required at very high cornering forces, and/or when you 'lift off' too quickly. In most situations the enforced transfer of weight from rear to front will simply reduce understeer. The secret with the 911 is anticipation, accurate judgement of speed and distance, correct positioning and 'finesse' with the controls.

Perhaps potentially, the most dangerous combination of inexpert (that is being polite!) use of the controls is to add harsh braking in a corner to the rapid and harsh reduction of power. That, naturally, merely compounds the problem and will, at high lateral 'g' forces, lead to very difficult situations. But you will only feel that happening, if you are driving so excessively fast, failing to match speed with vision, that you certainly would not want to meet yourself coming the other way!

The second characteristic which you need to be aware of is that under severe braking, because of the basic 42% front/58% rear weight distribution, the car will probably lock up the front wheels slightly earlier than the rear wheels; but only under very severe braking. Additionally, any slight deviation from straight line braking, under harsh application, will obviously tend to induce cornering forces on the already unweighted rear wheels.

Nevertheless it must be emphasised that these characteristics of the current 911 models are unlikely to be realised in any but the most extreme situations. Furthermore, the particular weight distribution of this car will, when driven in the manner we describe, produce very positive advantages of traction and manoeuvrability that in our view more than outweigh the potential minor disadvantages. Driven expertly and with understanding, the 911 will produce a superb combination of pleasure and practicality, unrivalled in any other car. In our experience, in the hands of a skilled driver, this race bred motor-car offers truly unrivalled margins of safety, and tremendous driving pleasure.

However, even the superb traction of the 911 will not save you in a third combination of ineptitude, should you get into it. The car's tremendous power, and the rapid response of the six-cylinder engine to the application of the throttle, can lead the unwary into the third pitfall. Imagine the tailgater is about to overtake a vehicle in a power driven swoop; the manoeuvre we described in Chapter 3. The road is wet with rain and autumn leaves. The overtaking manoeuvre is started from the nearside in a curving path, and as power is applied the car is progressively, and forcibly, unbalanced. This is dangerous.

The combination of a curving path and harsh acceleration to overtake, forces the car's weight on to the offside wheels. Then, as the car passes the lorry and the driver backs off the power quickly to slot in, perhaps even brakes, the car is unbalanced further. In both situations, at the beginning and end of the overtaking manoeuvre, the car is unstable and could simply spin into the path of either oncoming traffic, or of the overtaken vehicle.

These are all extreme situations where the inept or inexperienced (or both) driver is simply failing to appreciate the characteristics of the car. And if on the public highway, because of the car's very high handling abilities, would most probably be driving recklessly.

Perhaps the best way to experience the particular characteristics of the 911 is to describe taking the Cabriolet through one of the Porsche Customer Driving Day tests that provide an excellent opportunity to examine the car's manoeuvrability. The most appropriate test is usually set up on a corner of a race track, with a wide run off each side!

The cones are set out so that you have a set braking point from 50mph, a short distance to take off speed, then directly across your path a line of cones representing an obstacle with another line on the offside representing an on-coming vehicle, which together form a very tight chicane; right and then left.

Approaching the braking point at 50 mph, you must hit the brakes and lock-up, immediately cadence brake, taking off speed rapidly, then off the brakes, steer right, round the first line, cadence brake (not too hard or you'll spin) and then steer sharp left around the last line. Unless you take off enough speed in a straight line you will oversteer, and possibly spin, into the first obstacle. If it is wet you have to cadence brake, harder at first, in order to take off enough speed; and don't forget potential front-wheel lock up! If the car does start to oversteer, catch it instantly with opposite lock, or you will spin; don't brake.

In such extreme manoeuvring, because of the transfer of weight that takes place, often with sufficient force to reduce grip on the inside wheel to the extent that it loses traction and spins, the limited slip differential is a tremendous advantage. The power is immediately transferred from the spinning wheel to the wheel with the most grip; restoring traction. This has the advantage of extending the limits of the car even further, and ensures that you always sustain maximum traction and grip at the rear wheels.

This is a remarkably stable car, both aerodynamically and as far as the suspension is concerned. The 911 is exceptionally nimble, the brakes are absolutely outstanding and the handling predictable, up to and beyond limits that are never likely to be reached on the public road. However, even beyond the car's extraordinary cornering ability its behaviour is predictable, but more difficult to control. If you drive this car with the expert skills we describe in this book, assessing the speed you require for a corner or a hazard, braking and changing gear early, in a straight line, acting with responsibility and consideration, the car will reward you with immense driving pleasure. The car will simply not reward inept stupidity, with understanding. But, the limits at which such things will occur are very high and, in our view, are beyond almost any other car's abilities.

Optional Equipment

Whilst we have discussed the advantages and disadvantages of some of the optional equipment available in the preceding pages, we have summarised them below. Porsche has never provided 'add-ons', except where they add to driver/passenger comfort or enhance the vehicle's handling. The following equipment is available as optional extras on your 911 Carrera Coupé, Targa, or Cabriolet and will influence the handling, enhance the already extraordinary road holding, or improve your comfort. This is not an exhaustive list, simply the main items.

Limited Slip Differential

This is a friction device operating between the rear driving wheels, which operates in extreme cornering situations where the inside wheel will revolve at lower speeds than the outer wheel. With the normal differential, power is split between the rear wheels on a 50/50 basis. Under high cornering forces, the car tilts and weight is transferred to the outside wheels, increasing grip on those wheels but reducing grip on the inside wheel. This tendency is, of course, exacerbated if you brake or lift off the power harshly.

The limited slip differential will, the instant the inside wheel (or the outside, where that has less adhesion) starts to spin, transfer excess power to the rear wheel with maximum traction. The effect is to reduce the tendency for wheel spin under extreme conditions and provide more stable cornering. Particularly useful in the wet, or on ice or snow, this option is a major advantage for drivers who constantly drive in such weather conditions.

There are some disadvantages of using the limited slip differential, although we consider them to be only marginal disadvantages. There are occasions when you will notice the 'clicking' noise of the limited slip differential, when you are coming round a very sharp, very low speed turn or when parking. In dry conditions, you may detect a reluctance for the car to turn in and the car will have an increased tendency to understeer. Furthermore, the limited slip differential is subject to a higher degree of wear since its function is dependent on friction. These are the disadvantages, and are more than outweighed by the greater traction it gives you in adverse weather conditions.

SPORT SHOCK ABSORBERS

One of the many ways an engineer can influence a car's handling is to 'stiffen' the suspension. This reduces roll in cornering and therefore the tendency to reduce the weight (and grip) on the inside wheels. The use of Sport shock absorbers will provide this effect and reduce roll. They will also produce a slightly harsher ride.

If you have any intention of competition driving, it is certainly necessary to consider the use of this particular option.

WIDER ROAD WHEELS AND ULTRA LOW PROFILE TYRES

Another means of altering a car's handling characteristics is to alter the tyres or track width. Literally putting more rubber in contact with the road (tyre width) or, simply, increasing the width between the wheels (track width). Given a particular car, of any layout, if you reduce the rear track you will increase the tendency to oversteer, particularly if this is achieved by reducing the rear tyre section (if you use narrower wheels).

Porsche 911 Club Sport. The rear spoiler gives remarkable straight line stability on this and other 911 models.

Optional equipment of wider wheels and ultra-low profile tyres is available. The effects on handling are increased grip and a further extension of the 'limit' at which oversteer will occur.

The Weissach test drivers consider that tyres are major factors in the improved handling of motorcars. It is for this reason that they consider the minimum safe tread depth should be at least 3 mm across the whole tyre width. Their tests have shown a considerable reduction in tyre performance below that. This will apply particularly to the very wide ultra low profile tyres. In very extreme conditions with a great deal of water about, these tyres can tend to aquaplane more easily, if the tread pattern is less than this 3mm limit.

REAR SPOILER

The aerodynamic qualities of the 911 body shape does produce front and rear end lift but at very high road speeds. The effect of the rear spoiler is particularly important in high cross winds, rather than at high straight line speeds, since the effect of the spoiler is not felt much below 60mph (the national speed limit on two way roads) although above such speeds the reduction in rear lift is very considerable.

These optional extras will all improve the already superb handling of your Porsche 911 Carrera. However, the improvements will, in the case of the wheels, tyres and shock absorbers, be mainly sensed under extreme driving conditions.

Whether your car has standard equipment or some of the optional features we have described, the safety limits are very much yours to determine.

CHAPTER 9

Driving The Front-engined Transaxle Porsches

The New Generation

In its 40-year history of producing high performance road cars, Dr. Ing. h. c. F. Porsche AG of Stuttgart-Zuffenhausen has always followed a very clear design philosophy. From the first Porsche 356 to the current model range, Porsche high performance road cars have had only six basic body shells, and four basic engine layouts. The range of detailed development has certainly not been limited by its fundamental philosophy which demands excellence of basic design, a minimum model life of twelve years, sustained by a progressive and meticulous development programme. Indeed, the history of the 911 is a shining example of the success of such a policy.

It is, therefore, not surprising that the company debated long and hard before pursuing the new design concept of a front-engined, water-cooled engine, coupled with a Transaxle arrangement. As with many new departures, the development of the concept was not a coordinated or logical series of developments, but driven by somewhat varied imperatives.

In 1970, Porsche had developed the basic layout of its new cars. Porsche's development team had an acute awareness of market trends, and a concern about future legislation. It therefore decided that the first new model would have a 90° V8 water-cooled engine in the front, and a rear-mounted gearbox layout, for optimum weight distribution. The design team had examined more than 40 criteria including weight distribution and handling, noise, emission control, safety, interior space and performance, before coming to the front-engined, Transaxle solution, and this design layout was accepted in the Autumn of 1971, when work on the new car commenced. Furthermore, Volkswagen's joint sports car project (EA 425) provided an opportunity to try the new concepts on a smaller, less expensive and risky project.

By January, 1975, because of VW's fortunes (or misfortunes), Porsche had entirely taken over the EA 425 project. Details of the agreement are not important, but Porsche now had two new projects with a common design layout and both would benefit from the joint research and development effort. Another attraction was that Porsche would basically use an Audi engine, an Audi gearbox and a number of other VW-Audi parts, on the ex-Volkswagen project. The two cars now developed their very different market directions but their common engineering ancestry is obvious. This was the beginning of the new generation Porsches.

Porsche's aim is to produce superb 2+2 high performance road cars. The 911 is a classic example and the 'new generation' of Porsches are simply a modern reflection of that aim; true Porsches in every sense. The thinking behind the concept of a water-cooled front-engined car, with a Transaxle drive layout was dictated, not by fashion, but by reason, logic and a sure sense of the 'best' solution for a modern, high performance road car. In the same way that the 911 has proved to be a perfect solution of its time.

These new Porsches are of course very different cars. They range from the 924S with the Porsche 2.5-litre engine, to the 944 and the 944S, both with the same power unit but the latter having the four-valve per cylinder engine layout. Then there is the 944 Turbo, with a turbocharged engine producing 220bhp and giving this car astounding performance. Finally, there is the flagship of the Porsche range, the Porsche 928S series 4, with a V8 5-litre engine and many unique features, the only 'sports car' to win the Car of the Year award, which the 928 won in 1978. Although the whole range shares the same basic Transaxle layout, they are all very different, in performance, internal appointment, technical specification, and engine characteristics. Quite different in terms of performance and market, they do share a common set of handling characteristics, which only alter in a marginal way, from model to model.

Handling The Front-engined Transaxle Models

The new Porsches are all basically understeering high performance road cars with smooth predictable handling, and cover the whole Porsche range of performance. The front-engined Transaxle layout gives all of these cars an almost perfect 50% / 50% front to rear weight distribution. The Porsche 924S has a 52% / 48% front to rear weight distribution, the 944 Series are all 49% / 51%, apart from the 944 Turbo at an equal 50% / 50%, whilst the Porsche 928 Series are 51% / 49% front to rear. These almost perfect weight distributions give all of these cars a very high polar moment of inertia, and in each model the handling is predictable, the transition from understeer to oversteer smooth and easy to control, exemplifying the best in modern high performance road cars. Each model offers unrivalled smoothness of response to the application of power, under any conditions.

The aerodynamic design of each model has produced a low series of drag coefficients, with the 924S being marginally the better. The cars all have a servo-assisted steering system which has a progressive 'feel' built in. This is particularly important, since the steering forces

will not reduce but you will always have an amount of feel and the steering does not get progressively lighter as speed builds up, as it does with many cars.

Let us now consider how the particular handling characteristics of this series of Porsche models changes as we take a car around the same handling circle that we did with the 911. Naturally we have checked the car before we commence. The cars all portray the same basic understeering characteristics, but the 944 Series has the greatest tendency to understeer. As you start tracking round the circle, set the car in a steady state so that it is running round the circle at a steady, slow speed. As with all understeering cars, as you set the car up on the track and apply small increases of throttle (increasing lateral acceleration), you will find that you need slowly to increase the angle of deflection of the steering in order to remain precisely on the circle.

The understeer will build up, in comparison with a rear engine car, at a slightly higher rate in relation to a given circular velocity. Continue to keep the engine pulling the car and then, after the understeer has stabilised at a static slip angle, increase the acceleration very slightly once again. The effect will be to increase the slip angle. Naturally, this tendency for the understeer to increase can be achieved in two ways; either by increasing the rate of travel around a curve of a given radius, or by decreasing the radius of the circle. We are doing the former since it gives us a direct comparison of steering 'feel', and the slip angle, as we progressively increase the rate of rotation.

At Brands Hatch, the 1988 Model Year Limited Edition Porsche 944 Turbo with Sport Equipment, showing the Transaxle Porsche's understeering characteristics.

As you smoothly and progressively with no sudden movements of throttle or steering, build up the speed of rotation by increases of the throttle setting, both the slip angle and the lateral 'g' forces that you will feel pushing you to the outside of the seat will increase. Now, let us get to a point where you decide to stop the increase in the lateral acceleration. The degree of slip angle is too high and you wish to reduce it.

Therefore, instead of continuing to increase the throttle setting you lift off the throttle. If you do not alter the steering deflection the car will simply 'tuck in' to the circle and reduce the radius of rotation. Even at very high slip angles the car will not show any tendency to transfer from understeer to oversteer. The basic stability of the chassis and the almost perfect front to rear weight distribution will tend to ensure that grip is retained on the rear wheels, and the result is simply a reduction in the radius of cornering.

Whilst the result is different from that with a car with a rear weight bias your reaction, if you are to maintain the same turning radius, is the same – you reduce the angle of deflection of the steering. You have removed the speed, the cause of the problem, but to maintain your cornering path you have to counteract the physical reactions of the forces acting on the car.

Even if the road conditions are wet or slippery, the understeering characteristics of these models are so strong that to induce oversteer is simply to induce a gradual and entirely predictable and easily felt transition from understeer to oversteer. The cars are, if anything, so stable that the driving characteristics for swift road driving are always to maintain the balance of the car, rather than ever trying to induce oversteer; it is too difficult and you would be driving irresponsibly. Furthermore, it produces a slower and less smooth pace of travel.

In spite of this extraordinarily high degree of inherent stability the cars still have a great amount of feel. You can sense the balance, 'feel' the steering and control the car, even at very high slip angles. With all of the front-engined, Transaxle cars, it is extremely difficult, except perhaps under racing conditions, to induce them to oversteer. The degree of forgiveness of these cars is really quite exceptional. Coupled with the advantage of anti-lock braking systems, they must be the safest and most predictable high performance road cars available.

Finally we conducted some braking and handling tests at Brands Hatch, using a Porsche 944S with standard brakes and a Porsche 944 Turbo with Sport Equipment which had ABS braking. The test was very similar to that which you would do on a Porsche Customer Driving Day. A pair of cones acted as the braking point and each car was driven through the cones at a speed of 60mph. The 944S was driven by John Lyon and ordinary braking procedure used; he just locked everything up and kept the car straight. Then the same car was taken through the same test, using cadence braking technique. Then for the last test we used the 944 Turbo fitted with ABS brakes. The results were quite conclusive and the differences were as follows. The ABS car stopped 11.2 metres short of the 944S when it was simply braked to a halt with the wheels locked, and 4.8 metres short of when it was cadence braked. ABS braking systems definitely work.

Driving The 944 Series

In this case we will take the Porsche 944S. It is the current 16-valve version of the 944 and a car with all the handling and willing performance that epitomises Porsche's 'new generation'. The car's shape gives excellent stability and controlled power which is merely reinforced by the wheel arches and wide, low-profile tyres. The 16-valve engine of 2.5 litres produces 190 bhp at 6000 rpm, accelerates from 0-60 in officially 7.5 seconds (according to *Autocar* 6.7 seconds), and has a top speed of 142 mph. The main difference between the 16-valve engine, and the 944 8-valve engine is in the potential of the car's performance in the hands of an expert driver, and in the fact that the car offers the traditional Porsche 'S' advantages in the range. Better performance, better handling, and better top-end power.

The car is almost identical to the 944, with the addition of the double overhead camshaft. The automatic gearbox is not available as an option and the car is slightly heavier than the 944 at 1280 kg (2816 lbs), although power output and the maximum torque available are both considerably better. The Porsche 944S we have chosen comes with the standard braking system (although ABS is offered as an option) and the standard equipment all round. The car has, of course, the advantage of the TOP cylinder head design and the knock-sensing Bosch Motronic digital engine management system.

Approaching the car from the rear, check the rear light cluster, and that the exit behind is clear. When you are sitting comfortably, adjust the external and internal mirrors. The passenger is settled, doors closed and seat belts on. Switch the ignition on and the engine fires naturally and instantly. Check gauges and warning lights. The cockpit ergonomics are certainly 'right', although the handbrake is, perhaps, a little too far away.

On the road, the efficiency of the chassis design allows some exuberance to be safely used on dry sweeping bends. The balance and grip of the car rapidly make themselves apparent, as does the need to keep the engine speed up, to get the best from the superb torque of the engine. A cross-country route is a joy in this car, as the superb road holding, general handling, and braking characteristics make safe, fast and effortless progress easy. The chassis design makes it practically impossible to beat the road holding in the dry, under normal road driving. The servo-assisted rack and pinion steering is a perfect combination of feel and assistance, particularly when powering the car through sweeping bends. Under hard cornering,

the car tends towards mild and predictable understeer; oversteer appears to be almost impossible under normal road conditions. You certainly cannot induce it in the dry without ignoring the basic rules of advanced driving. The overall impression after a testing run is further confirmation of the excellence of the chassis, and the perfection and marvellous balance of the car.

The 944S covers miles in a tireless and very quiet style – the latter being a very important factor in long distance touring. In a typical English drizzle one is grateful for the excellent ventilation system which is fitted as standard. Simple to operate and very efficient, it clears the screen and side windows inside as quickly and efficiently as the front and rear wipers act on the outside. The controls for electronic air temperature and flow work perfectly. All round visibility is very good and the three mirrors (wing mirrors electrically operated) provide a complete view of the disappearing road. The level of road noise from the wide, low profile tyres can be obtrusive on some surfaces.

In our drives, we put the 944S through a few tests. Whether on dry, wet or muddy country lanes, the car never gave us a moment of concern. The superb seats proved their worth, always comfortable and offering the level of support needed in a car with such exceptional cornering performance. Several simulated emergency stops convinced us that the braking system is not just good, it is up to the car. However, in the unpredictable and often muddy country roads the optional ABS system would certainly have added a welcome degree of extra safety.

Overall impressions of the car are of a chassis and suspension that are capable of absorbing yet more power. In the 944 Turbo they are both put to the test, of course. The handling characteristics of the car and of its 'feel' are very different from the 911 Carrera Cabriolet. The Cabriolet is very lively, whereas the 944S has a very stable feel about it. In spite of the sometimes obtrusive road noise, it is still an extraordinarily relaxing car to drive, and daily mileages of over 350 miles, would certainly not, in our view, be too onerous.

The car is almost too efficient in its capacity to cover large mileages with a degree of certainty and sure-footedness that are commendable, if less exciting than other models. The plain facts are that you will never (a high risk word in this book) outwit the handling of the car either in the wet or the dry, if you are driving with care and consideration for other road users, and matching speed with vision. The car has so much primary safety 'in hand' that it is exceptionally difficult to catch the chassis out, with the power available. Handling, even on wet muddy roads is always predictable, and when you do induce oversteer, and it was only possible on muddy roads, it can be readily and easily corrected. This car is exceptionally well mannered and predictable and yet has enough performance to make quick, safe, and very comfortable continental touring a real pleasure. The steering

Opposite above The Porsche 944S.

Opposite below Porsche 944S cockpit.

is very positive and light, both on the arms and the nerves, and the servo-assistance is progressive and never intrudes into your 'feel' of the car.

The front-engined, Transaxle design really needs no greater justification than the sheer pleasure that the car can provide to the expert driver. This car has, probably, only one major failing and that is it could well induce a feeling of vast over-confidence in the inept or inexperienced driver. It is so tractable and its handling is so predictable that, given care, you will never be in a position to exceed its handling abilities in normal, or even abnormal, road driving; providing you use judgement. The handling of the Porsche 944S, even under pressure, remains predictable and exceptionally civilised. Fast entry into a corner brings on gradually increasing understeer, without any lightness of the steering; so well and progressively does the power-assisted steering reduce its contribution with speed. The rear disc brakes have a 'force regulator' which prevents the rear wheels locking under severe braking, in effect making it very difficult to create oversteer. Even if you manage to induce oversteer, intentionally or unintentionally, the 49% / 51% front to rear weight ratio ensures that the transition will be gradual, predictable, and controllable. The car is so well balanced that even under relatively high cornering forces, if understeer is excessive, you can back off the power to reduce it. The car will never snap into oversteer but gradually rear slip angles will increase, front slip angles reduce and the car will slide quite controllably and easily.

The front-engined, Transaxle cars are capable of lapping the 'mountain course' at Weissach quickly, and only 3.5 seconds slower than some of the quicker 911 models, with nearly 30% more power. The 944S has these qualities in abundance, being quicker through all of the intermediate performance speeds than either the 944 or the 924S. At engine speeds above 3000rpm, the car really comes into its own, although at 3100rpm in fifth gear you will be at the national motorway speed limit! The important advantage is that the power available for overtaking, which is impressive and flexible, always appears to be available.

The car is an ideal touring car, capable of relaxed high speed cross-country motoring. Noise levels, apart from wheel/tyre noise on some road surfaces, are very low, with wind noise at a minimum. Your 944S is a well balanced, beautifully built driving machine which will ensure you travel quickly, safely, and with some positive driving effort, smoothly and unobtrusively.

But what of the other models in the range? Let us briefly examine the handling characteristics of the two extremes of the spectrum.

Driving The 928S Series 4

This car is one for which it is very difficult not to invent a whole new range of superlatives. So let us simply stick to the fundamentals of driving this remarkable modern Grand Touring car. Technically, the car is at the forefront of automotive engineering. The 32-valve, V8 engine develops 320bhp at 6000rpm, and has all the smoothness and response which you can not only expect from a V8, but imagine. The engine produces over 300lbs/ft of torque between 3000rpm and 5500rpm, with a maximum of 317.1lbs/ft at 3000rpm, with a wide torque band above 3000 rpm, which make it one of the very fastest production cars in the world. Performance figures are simply a top speed of 167mph, a 0-60mph time of less than 5.8 seconds, and similar acceleration performance through the speed range.

The external appearance of the car is very smooth and through aerodynamic development the Cd value has been reduced to 0.34, with the front intakes closed. The result is a car which is stable throughout its considerable speed range. Interior noise levels are exceptionally low at even the highest speeds. This is Porsche's first V8 and the central spark plugs, TOP cylinder head design, and a superbly balanced engine make it a very smooth unit indeed, even if you ever get the opportunity to tax its power output; which will only happen on the track or autobahn (if then!).

Technically the car is unlike anything Porsche, or any other manufacturer for that matter, produces, with many unique features.

The Porsche 928S series 4.

Porsche 928S series 4 cockpit.

The original 928 won the 'Car of the Year' award in 1978, and it is not difficult to understand why it was the first sports car ever to win the award. Approaching the car, its smooth lines are immediately obvious, from any angle. The front and rear bumpers (they are not immediately recognisable as such), are made from lightweight, impact resistant polyurethane and form a smooth and completely uncluttered front and rear design. The front fog lights are set into the bumper, as are the front and rear indicators. The interior is spacious and very well appointed, and both front seats have full electrical adjustment. Just in case you were foolish enough to let anyone else drive your 928S series 4, the seat adjustment has a memory system which covers the external mirrors.

There are a number of features of this car that we need to describe before our drive. Firstly, ABS brakes are fitted as standard, and the brakes are exceptional; they will pull the car from 62.5mph to zero in 3.3 seconds. The manual, five-speed gearbox has an unusual but highly logical layout; reverse is left and forward, first is back and left, second is forward, third back, fourth and fifth are to the right forward and back respectively. In spite of the tremendous power and torque, the short gear lever makes changes smooth and easy. The car has a unique suspension system, in addition to the Transaxle layout. The front suspension is more traditional with double wishbones, coil springs and telescopic dampers. The rear suspension has the "Weissach Axle" which reduces any tendency for oversteer, by compensating for any toe-out action of the rear wheels if you lift-off

The Porsche 928S series 4.

the power in a corner. Additionally, this axle will reduce the lifting of the front of the car under extreme acceleration, and therefore maintain front wheel grip.

The car's handling is, under all normal road conditions, totally neutral or, at the most extreme, marginally oversteering. The limits on this car's handling will, in our opinion, never be reached in normal road driving unless in the extremes of ice or snow. Even then the car has the ability to satisfy the most demanding driver. Even in the wet, harsh use of the throttle will only produce oversteer if the enormous power of the engine is used indiscriminately, and with a total lack of response to the car's 'feel' that will be giving messages in triplicate through the steering wheel and chassis. When it does occur, at very high cornering forces, the breakaway is slow and progressive, and can be instantly and safely corrected by lifting off the power and gently correcting the steering. The action of the Weissach axle dampens any tendency to oversteer so effectively, that there is practically no chance of getting into a tail wagging swerve; well, unless you should not be driving at all.

Under very hard driving you can make the car tighten its line through a corner by lifting off the power, but in our experience the Weissach axle will cope with any tendency for the rear end to even think of coming round. In the wet it is, of course, very easy to induce wheelspin, but the massive front and rear tyres put down such a large amount of rubber that even the very heavy footed have plenty of warning, and time to correct. The impression is not of driving a big

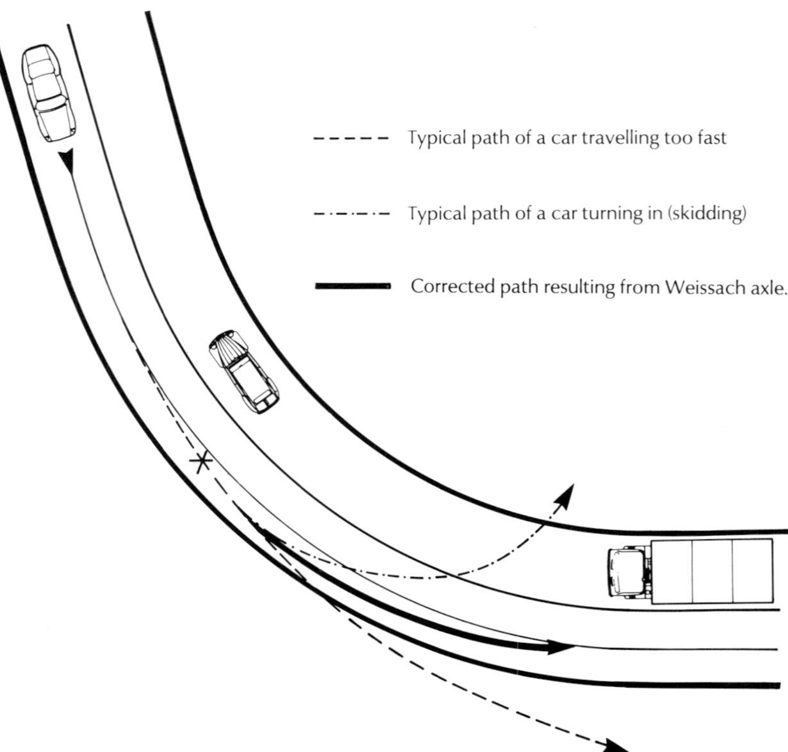

The Weissach axle neutralises handling.

Grand Tourer at all since the steering, brakes and power are so capable, they make the whole car feel like a very taut and physically much smaller high performance car. This car can be driven immensely quickly and yet with complete assurance, it is so well balanced.

The power assisted, rack and pinion steering is beyond praise and has all the 'feel' required of such a car, never giving anything less than superb feedback. Even on the track the steering is light and positive, giving the opportunity to be accurate, positioning the car into corners with precision. The ABS system will bring the car to a very rapid stop in the worst conditions.

The optional extras seem to be all on the car in standard specification although the addition of a limited slip differential would add to the ability of the car to put the power down to the rear wheels in extreme situations. Indeed, with the effectiveness of the standard suspension system, the conditions would have to be very extreme!

In some ways this car defies description because you simply have so little to compare it with. It is a fundamentally different car from the others in the range and has active safety limits that will be beyond the expectations of normal road use, in any circumstances. The power and grip available, mean that motoring is a matter of your limits rather than the car's.

Driving The 924S

The 924S has very similar handling characteristics to the standard 944, but it is a very different car in terms of appearance and in certain handling situations. Technically the cars are similar and produce the same power output when fitted with emission control systems. The 924S is powered by the 2.5-litre, four-cylinder, eight-valve engine producing 160bhp at 5900rpm. Maximum torque of 155lb/ft is produced at 4500rpm, giving a wide ranging, smooth and quick response to power application, helped by the car's slightly lower weight at 1240kg. The Cd value of 0.33 is the lowest of the range, and gives the car very good straight line stability and low wind noise, throughout its speed range.

You have completed all of your checks and we are ready to go. On the road the car immediately impresses with its stability and responsiveness. But what distinguishes its handling from the other models in the range? Firstly, the car does not understeer as much as the 944S or the 944. The initial understeer occurs in very similar style but the slightly narrower tyres and track will reduce the cornering force at which understeer will first occur. However, the understeer is not as pronounced and is very predictable and comfortable. The tuck-in which happens as you back off the power happens over a wider band of speed and cornering forces, and is slightly less firm. To induce oversteer in normal road driving is very difficult, and when it does happen it is very predictable and controllable.

The car seems less brisk than the 944, but nevertheless actual acceleration times are very creditable, with a 0-60 time of less than 8.3 seconds. The servo assisted rack and pinion steering on later models is light and positive at all speeds and manoeuvring at low speeds is a comfortable operation.

As with all the cars in the range it is across country that the 924S comes into its own, the light, positive steering giving excellent 'feel' and the responsive and predictable chassis enabling you to take the car to the limit of safe speed and vision, and still be a long way from the car's handling or performance limits.

The car is always totally predictable with a complete lack of nervousness, although of all the optional equipment available the Sport shock absorbers and the thicker front and rear anti-roll bars would certainly extend the already high handling limits. In particular they will reduce the, admittedly low, level of roll under very hard cornering, giving a flatter cornering posture, with additional grip in adverse conditions. The Sport seats would be a personal choice to complete the package.

This car gives a great deal of pleasure and rewards the attentive and skilful driver with an easy, relaxed and very smooth drive, at a quick and satisfying pace. As in all well balanced high performance road cars the brakes and steering inspire a confidence that is never betrayed. The surefooted and predictable chassis and excellent torque of the engine offer an immensely safe, swift and satisfying combination.

Optional Equipment

The range of optional equipment available for the 944S and other models in the 2.5-litre range is extensive. Even in standard specification these cars have quite extraordinary active safety limits of road holding, handling, and an ability to cope with emergencies. Although we have described the effect of some of these options in the text, the following is a brief summary of some of the main items of optional equipment available. The following additions to the basic specifications will all enhance the exceptional handling and performance of these cars.

The only option applicable to the 928S series 4 is the limited slip differential.

Limited Slip Differential

The value of this option is certainly felt in extreme conditions of snow, ice or in the wet. The cars' superb handling reduces the need for this option, except in extreme conditions. However, there is no doubt that it will extend the already impressive limits of the cars' cornering power.

The disadvantages were outlined earlier in Chapter 8 but if you are going to fit any of the suspension options it becomes a must.

ABS Braking System

The standard equipment on the 944S includes a pressure regulator which reduces the tendency for the rear wheels to lock-up under severe braking. Nevertheless, in our view, the addition of ABS brakes will increase the safety reserves of the car, no matter who is driving it.

Practical proof of this is in the brake tests we carried out at Brands Hatch, between a 944S with standard disc brakes and a 944 Turbo with Sport Equipment which had ABS brakes as standard. John Lyon stopped the 944 Turbo with Sport Equipment 4.8 metres short of the 944S when he cadence braked, and 11.2 metres short when he simply 'locked up' the brakes; stopping from 60 mph in each case.

The weather conditions on the day of our tests were fine and dry. We are convinced that, in the wet, the differences between stopping distances would be considerably greater. There is no doubt that ABS braking systems stop you more quickly and safely.

In approximate terms, the difference in braking distances would mean stopping before an obstacle with ABS. Hitting it at about 12

The advantage of ABS brakes. The 1988 Model Year Limited Edition Porsche 944 Turbo with Sport Equipment has ABS brakes as standard equipment, and it stopped 11.2 metres short of the 944S.

mph after cadence braking. Hitting it at about 18 mph by simply locking up the wheels. At either speed you will suffer, and so might someone else!

ABS brakes stop a car more quickly and more consistently than any expert driver, but that is not their major advantage. You cannot steer when the steering wheels are locked, but *with the ABS system you can brake as hard as possible and still retain steering control. This advantage is probably the single most important safety feature of any of the optional extras available. You can use the maximum stopping power of the braking system and still steer away from a hazard.*

For the '89 Model Year, we understand ABS will be standard equipment on all 4-cylinder models, a move we heartily welcome especially in view of the fact that, with the exception of the 924S which will no longer be produced, all the models will become more powerful.

Wheels and Tyres

In spite of the fundamental stability of the 944S chassis, the addition of wider wheels and ultra-low profile tyres will offer yet more grip in all conditions. Additionally, the wider wheels and tyres will increase the available cornering force and reduce, further, any oversteering tendency in extreme conditions.

Although these tyres may tend to be less stable in very extreme weather conditions than narrower tyres, the difference, if any, is very marginal. Furthermore, if you maintain the tread limits set by Weissach, which is a minimum of 3mm of tread all over the tread pattern, the difference will be even less. The wider your tyres the more concerned you should be about tread depth in the winter.

Sport Shock Absorbers

The addition of this optional equipment will create a car that will corner with considerably less roll (although it does not exactly roll a great deal in standard form), and therefore retain more adhesion on the inside wheels.

In normal road conditions you will notice the slightly harsher ride probably more than the improved cornering. However, if you do a great deal of cross-country motoring or are planning any competition driving, then Sport shock absorbers are a useful option in the former case and essential in the latter.

Sport Seats

Personally, for reasons of support and comfort at higher cornering forces these seats are very desirable. Nevertheless it should not be assumed that the standard seats are in any way less than excellent. The Sport seats do give additional support on long journeys across country, and can contribute to a less tiring journey.

The choice of a particular specification for your car is a very personal choice and will depend upon your motoring needs. In our view, the biggest contribution to safety are the ABS brakes, and limited slip differential. If you want improved performance, then the Sport shock absorbers and wider wheels with ultra-low profile tyres should be your choice. However, the performance improvements will only be noticed at very high cornering forces, or in extreme weather conditions.

CHAPTER 10

TURBO EFFECT AND AUTOMATIC TRANSMISSION

TURBOCHARGING

Porsche's decision to launch the Type 930, or the 911 Turbo as it is more commonly known, at the Paris Motor Show in 1974, was considered by many to be a supreme act of faith, by others to be flying in the face of reality. The car was built in order to gain homologation for a Group 4 racing car and formed the basis of the 934 and 935 racing cars. However, Dr Führmann decided that the car had to be a truly luxurious Porsche built, not as an evolution model, but as a supercar to take the market lead. The competition cars became the evolution models, derived from the production road cars. But first, the Porsche engineers had to utilise their racing experience of turbocharging and fuel injection.

The crucial factor in Porsche's successful application of turbocharging to its road cars was the experience gained in developing the turbocharged version of the 917 racing car; the ultimate weapon. Technically, turbocharging is simply the process of using the engine's hot exhaust gases to drive a turbine that increases the engine's air intake; pushing air into the inlet manifold under pressure, regulated by a boost control unit. This also requires careful regulation to ensure optimum air/fuel ratios under all operating conditions. Turbocharging may be a simple system but its application is more difficult. There were two difficult problems; controlling the boost pressure, and improving the relationship between the fuel injection system and boost pressure, since fuel flow needs to increase as boost pressure increases. Bench testing alone would not provide the answers and so throughout 1972 and 1973 Porsche undertook an intensive and very long distance programme of track testing with the 917, and these problems were resolved, far more successfully than was thought possible.

Porsche's engineers had used a wastegate arrangement on the 917 to control boost pressure. But they finally solved the second part of the problem, relating fuel flow to boost pressure, through very careful re-programming of the electronic injection system.

As a direct result of this extensive racing programme, Porsche's engineers had resolved the problem of turbo lag, the time gap between depressing the throttle and the turbocharger building up sufficient

The turbocharged Porsche 917-30; race bred development reduced turbo lag and improved fuel injection.

air pressure to boost power. Porsche had also so improved the relationship between the fuel injection system and the level of boost applied to the engine that it now had a very responsive and tractable turbocharged system.

By 1973 the success of these developments in turbocharging had led Porsche to a number of conclusions. Firstly, the company realised that the turbocharger was now sufficiently flexible and responsive to be used in a road car. Secondly, although it realised that the fuel crisis meant that greater engine capacity was probably no longer acceptable, it still needed to push the 911 back to the forefront of production car racing. These two conclusions led Porsche to an imaginative reaction to the economic crisis of the early 1970s, and to the launch of a brilliantly innovative commercial success. The answer was the forerunner of all successful turbocharged racing engines, and gave Porsche a formidable string of racing victories that has yet to be beaten.

Its first step towards road going application was to turbocharge a Carrera RSR for the 1974 Le Mans, although race regulations (a 1.4 multiplier applied to turbocharged engine capacity) dictated that this car had a 'baby' 2.1-litre engine, to compete in the 3-litre class. Immediately following the Le Mans success, when the car came second overall, a similar turbocharger was fitted to the 3.0-litre flat-six production engine.

The combination proved to be tractable, extremely powerful and less noisy, as turbocharging is a quiet way of increasing power and needs a smaller tail silencer. The first Porsche 911 Turbo production models were sold in 1975, and demand proved so strong that rather than curtail production at 500, as was originally intended, production was increased and by 1988 over 13,000 had been sold. The decision that many thought to fly in the face of reason has proved to be one of Porsche's most successful.

The car's success was perhaps, inevitable, because in spite of its truly exceptional performance, development work with the 917 meant that the turbocharged 911 had lost none of the practicality of the naturally aspirated car. In 1977, when the engine was uprated to 3.3 litres, it became the first production car to be fitted with an intercooler. This unit not only improves performance but provides for greater engine longevity. Then in 1984, a standard Porsche 911 Turbo covered one kilometre from a standing start in 23.985 seconds to win the title "The Fastest Accelerating Production Car in the World", a title the car still holds.

The second of Porsche's road cars to be turbocharged had a similar but later background. In 1981, Porsche entered Le Mans with a Transaxle car, powered by a 2.5-litre turbocharged water-cooled engine. Although enthusiasts had to wait until February, 1985, for the first production car, it was an obvious development for more expansionist times than the early 1970s. Porsche had resolved the emission problems, and the 944 Turbo was the first Porsche to produce identical power and performance with or without exhaust gas equipment, and to run on unleaded petrol down to 95 octane.

The 944 Turbo uses a turbocharger with a water-cooled jacket. This unit is fitted to the induction side of the engine to reduce thermal loads and its positioning offers the added advantage of a shorter run to the inlet manifold thus further reducing turbo lag. Other changes to the 944 2.5-litre engine were thicker cylinder walls and a lower 8.1 compression ratio. The system also had an air-air intercooler. Porsche used ceramic material on the exhaust ports to improve the performance of the catalytic converter (when fitted).

A range of other modifications to the standard 944 specification was necessary, including strengthening the gearbox, uprating the brakes and stiffening the suspension. Body changes included a new front spoiler and a rear undertray which, with the flush front screen reduced the Cd to 0.33 and helped crosswind stability; never really a problem. The performance of the 944 Turbo put it straight into the 'Supercar' league, with an extremely relaxed and effortless top speed of 152 mph, and a 0-60 time of less than 6.0 seconds.

So what are these two cars like to drive? Their handling characteristics are not fundamentally different from their non-turbocharged cousins, and we will not repeat what we have said. In each case we will concentrate upon the slight differences in handling and the appropriate driving techniques.

DRIVING THE 911 TURBO

The instant you turn the ignition there is a different, quieter sound from the 'flat' six-cylinder engine. Check the gauges for oil contents, oil pressure; the former will only read full after the engine has reached operating temperature, but don't forget to check the oil level when the car is idling on level ground. The oil pressure should always read above 3.5 bar at 5000 rpm.

As we drive off the car immediately feels very much stiffer and the ride firmer than the standard 911 Carrera Coupé, with the 3.3-litre engine singing, quietly, behind us. This car produces its torque in a peaked band of 318 lbs/ft at 4,000 rpm, but still manages to produce over 271 lbs/ft between 3,000 and 6,500 rpm, below 3,000 rpm the torque tapers off quite rapidly. Peak power of 300 bhp is produced at 5500 rpm so, for performance, the engine speed has to be sustained at above 3,300 rpm, with the boost pressure rising to its maximum of 0.8 bar. Unfortunately, if you allow the latter to happen without great self control you will be in danger of being over the national speed limit in 6.0 seconds, and before you have reached 4,000 rpm in second gear.

The car's handling is essentially very safe indeed up to limits which you will never hope to reach in normal, responsible road use. On dry roads the car's handling is absolutely neutral at very high cornering forces. The wide ultra-low profile tyres putting very large prints on the ground at all four corners as the stiff suspension reduces roll to a minimum. Push the car *very* hard and understeer will set in, but considerably later than on the standard 911 chassis. Increased power will simply increase understeer; although you will have to watch the surge of power above 3800 rpm. The secret is to be light but progressive with the throttle, particularly through corners. Use your anticipation and judgement to set the car up in the appropriate position and a steady throttle setting to maintain the required speed through the corner. You can then use the power as vision opens up and you exit the corner, matching speed with vision.

If you do power yourself into a corner too fast, understeer will increase and as you lift off the power, decreasing the understeer, you must instantly but gently correct the steering. However, the tendency to oversteer is significantly reduced by the wide tyres and the tremendous grip that they give on all surfaces. The handling limits on this motorcar are, frankly, well beyond the limits that most drivers would begin to reach. Even when driving the car on a track it is incredibly difficult to induce oversteer; much more difficult than on a standard Porsche 911 Carrera. Because of the slightly greater rear weight bias the steering is perceptibly lighter than on the standard Carrera. However, the wider tyres and the suspension increase the front end traction to compensate. You really have to try very hard to make the car go in any direction except that in which you point it.

Road going application of racing car development, The Porsche 911 Turbo Cabriolet with Sport Equipment.

When overtaking, early positioning is essential. Setting the car up for a smooth overtaking manoeuvre, you will then be in the correct position to accelerate in a straight line, so that as the engine speed rises above 3800 rpm you are settled, and not on a curved and possibly unstable accelerating path. In circumstances where you are in extreme weather conditions or using the car's exceptional power, the optional limited slip differential is an immensely practical addition to the car's handling performance. Although it may still produce the minor annoyance of noise in low speed manoeuvres, the advantages will be considerable.

The tremendous power of the turbocharged engine combined with the 911 Turbo running gear make for an extraordinarily safe, predictable and practical motorcar of exceptional performance. In the wet, aggressive use of the power will possibly overcome the enormous traction of the tyres, so in such conditions you should use the power with judgement and responsibility. The appropriate technique of feeding the car into the corner with early anticipation of speed and position, maintaining the appropriate road speed with the power, and not applying increased power until you exit the corner, becomes more critical. Any irresponsible use of the power and you will be 'hit' with the rapid increase of torque above 3,800 rpm, in the middle of a corner.

The brakes on this car, as with the rest of the specification, are more than capable of controlling the speed and acceleration. However, they are exceptionally powerful and therefore braking will produce very rapid deceleration. If used in a corner this will induce understeer and unweight the rear wheels. The answer is don't, unless you brush the brakes with a trail-brake technique. But never use this technique unless you are extremely skilled and have had expert instruction.

In spite of the front to rear weight balance of 38% / 62% the Porsche engineers have achieved a balanced and stable car that is very difficult to beat. Indeed, after John Lyon drove it around Cadwell Park, and on the road, his view was that you would not be able to reach the car's limits driving under normal road conditions, with the finesse, skill and expert judgement we have described.

DRIVING THE 944 TURBO

The 944 Turbo is externally different from the standard 944, and the difference becomes apparent as you approach the car. The rear undertray spoiler is immediately obvious and, as you go round to the front of the car, the neat front polyurethane panel gives the car a smooth streamlined appearance. The turbocharger and the related amendments made to the suspension and braking systems have given this car a perfect 50% / 50% front to rear weight distribution. The turbocharged 2.5-litre engine produces 220 bhp at 5800 rpm with maximum torque of 243 lbs/ft at 3500 rpm. However, torque is very high over a wide band with 150 lbs/ft at 2000 rpm, 225 lbs/ft at just over 3000 rpm and over 187 lbs/ft at 6000 rpm. Therefore, from about 2500 rpm you have a considerable torque increase, remarkable for a turbocharged engine.

The car's brakes and suspension have all been uprated to cope with the extra (60 bhp over the 944 and 30 bhp over the 944S) horsepower and torque. The main differences are that the suspension has independent MacPherson struts with coil springs and tubular stabilisers whilst the brakes have more powerful four-piston calipers, and ABS is standard.

On the road the car feels slightly heavier than the 944, although it is a feeling induced more by the wheels and tyres than all up weight. The car's inherent stability undoubtedly contributes to this sensation. Driving the 944 Turbo over a cross-country route was a revelation. The car is relaxing, effortless, but very swift. Pushing the car hard, the turbocharger comes in very smoothly and progressively, there is no sudden surge of power. Concentrating on the road ahead you wonder when the turbocharger will come in, until you sense that you are pressed very securely into your seat and the road ahead is approaching very rapidly. From rest this car will accelerate to 60mph in less than 6.0 seconds and you will still be in second gear. If you are on an autobahn, 100mph will come up in 15 seconds and the top speed is over 150mph. So, apart from the shattering performance what is the difference between this supercar and the other models in the 4-cylinder range? Not a lot is the answer.

The car understeers later and at much higher cornering forces than other models in the range. Indeed the perfect 50% / 50% front to rear weight distribution is immediately apparent, although the stiffer suspension undoubtedly maximises the benefits of this advantage. The more you push the car, subjecting it to increased cornering forces, the more the car understeers; gently at first, then at progressively increasing slip angles. In driving tests on a dry track at Brands Hatch, we found it very difficult to make the car behave in any other way, no matter how hard it was pushed. On our drive back to Reading even rough road surfaces did not upset the smoothness and stability of the car; it simply emphasised the road noise from the wide tyres. Even though the limited slip differential option will only be noticed in extreme situations it is an option that we would suggest will give this car an additional safety factor. With the tremendous power available if the grip under either rear tyres alters, the ability to transfer power to the wheel with the greatest grip will give an additional safety margin.

If you are travelling quickly through a bend and you have to lift off the power, for reasons of safety or vision, the understeer does not snap off, the car just settles further into the corner and the steering can be easily corrected. Pushing the car into corners harder and faster you can induce a gentle and easily sensed transition into mild oversteer, at which point you will be travelling very quickly.

The current 944 Turbo is definitely in the supercar class, by anyone's standards. It will do everything that almost any other car in its class will do, but with absolutely no fuss or tantrums, in a very relaxed and civilised manner. Even in terms of straight line speed and stability this car will only give second best to either a 911 Turbo or the 928S series 4. Overall a very favourite Porsche and, almost, the most relaxing at very high road speeds.

Automatic Transmission

There are currently three models available with automatic transmission, the 924S, 944 and the 928S series 4. The addition of automatic transmission does not, in fact, alter the handling characteristics of any of the cars in any way at all. The difference is primarily in the manner in which you should use their respective automatic transmissions to ensure swift, safe and sure progress. In Chapter 1 (see pages 47-49), we have already described the basic functions of automatic transmission, and the principles of using this fatigue reducing instrument in a high performance road car. Therefore, we will briefly, describe the particular automatic transmission fitted to the 924S and 944 Automatics, and the 928S series 4 Automatic and then remind you of the important techniques applicable to handling.

The automatic transmission used on both the 924S and the 944 is a three-speed box, whilst on the 928S series 4 the very much greater

Porsche 928S series 4 automatic.

power and torque requires a four-speed automatic gearbox. The mechanical principles of both are very similar with the Bosch engine management system on the 8-cylinder unit retarding the ignition by 350 milliseconds during upward gear changes to enhance gear changes and ensure greater gearbox longevity. With an automatic gearbox, that most abused component, the clutch, is replaced by a hydraulic torque converter and fluid drive. The converter seals can only be damaged by very serious misuse and lubrication is less frequent. In all a very convenient package and if used with skill and acceleration sense, it will detract little from normal road performance.

The engine can only be started in 'P' Park, or 'N' Neutral. Once you have started the engine, apply the footbrake before you select 'D' Drive, to prevent any creep forward. Incidentally, idling speed on the 928S series 4 is reduced when 'D' is selected to minimise this very possibility. Similarly, apply the footbrake before you select 'R' Reverse.

The best results from an automatic gearbox can only be obtained if you understand the basic influence of the torque converter. In normal circumstances the automatic gearbox will provide extremely rapid take-off, as the throttle is depressed to the point before the kick-down which automatically changes down one gear. The initial acceleration can then be transformed as you depress the accelerator pedal the last inch to employ full kick-down. You should increase pressure gradually and progressively, only flicking the foot over for the final inch to engage the kick-down.

You should guard against using kick-down when cornering in the wet, ensure that such acceleration is only used when the car is straight. The control available with the automatic gearbox is, providing it is used properly, quite exceptional in each case.

When cornering through open bends on a dry road it is important to remember the old adage 'in slow out fast'. In every corner it is vital that you assess speed correctly, so that you allow the car to settle and keep the engine pulling the car through the bend in the appropriate gear, without using the kick-down or allowing the car to coast. Then you should use the kick-down at the start of the expanding radius steering path, sustaining the power progressively thereafter. In corners where the line of sight is restricted you should not use the kick-down until after the line of sight opens up.

It is important to remember that for all normal motoring the automatic gearbox should be used as it is meant to be used, automatically. In order to drive swiftly and safely, using the full performance of the car you will need as much acceleration sense as you would with a manual gearbox. In all normal circumstances you should be in 'Drive' for all forward movement. You will remember that in Chapter 1 we gave you three distinct and separate situations when control can be improved by the manual selection of an intermediate gear. These are as follows:

1. To control road speed through and between a series of bends, provided that the anticipated speeds are within the range of road speed in the intermediate gear selected.
2. To help the braking system to hold the car back down a long steep hill. Brake first to a safe descent speed, which is invariably very slow, at the top of the hill and then select second intermediate.
3. You may choose to select an intermediate gear to maintain position in a stream of traffic, thus preventing excessive use of the brakes.

In conclusion, the use of automatic transmission detracts very little from these cars' normal road performance. Moreover the convenience and lack of wear and tear are very conspicuous advantages for many drivers.

THE 928S SERIES 4 AUTOMATIC

The cockpit interior is distinguished only by the 'T' shaped automatic gear lever and the illuminated selector indicator in the bottom quadrant of the rev. counter which tells you which gear you have selected. The gearbox has 'P' Park, 'R' Reverse, 'N' Neutral, 'D' Drive, '3' and '2' intermediate gears marked. The important speed ranges through the gears are as follows; 1st 0-36mph, 2nd 0-70mph, 3rd

19-118mph and top 27-164mph (an academic 3mph less than the manual).

It is interesting to note that the 928S series 4 will, under certain conditions, move off from rest in second gear. This feature enhances driver comfort and provides additional fuel savings.

The exceptional power and torque of this car make it admirably suited to the very flexible automatic gearbox, although there are marginal differences in the intermediate acceleration times; the 0-62.5mph time is 6.3 seconds. The others are quite quick enough, since you will be at the legal maximum in second gear. The power is always available and acceleration for overtaking can be applied with confidence and absolute safety, given you use judgement and anticipation.

The addition of the limited slip differential is a positive advantage with automatic transmission. It can offer additional traction in extreme situations such as ice and snow.

In traffic or for long distance cruising the automatic gearbox really does come into its own. In this car the superb engine and advanced suspension systems make the sacrifice in terms of performance, an extremely marginal one, and a sacrifice on which each individual driver has to decide.

The 944 Automatic

As usual, the 'T' bar automatic gear selector replaces the gear lever and has all of the positions indicated along the right-hand side of the centre console. In the bottom quadrant of the rev. counter the positions of the automatic gearbox, 'P' Park, 'R' Reverse, 'N' Neutral, 'D' Drive, '2' Second and '1' First are displayed in a descending curve. An illuminated circle moves round the signs and tells you which gear you have selected at a glance.

The top speed of 137mph is identical to the manual version of the car, although the intermediate acceleration times are marginally slower. The 0-60mph time is 9.6 seconds and the speed ranges through the gears are; 1st 0-55mph, 2nd 15-93mph and 3rd 25-137mph.

The car is extremely relaxing and comfortable to drive and, as we have said, handling is not affected in any way. Indeed, at Cadwell Park we tried several of the Porsche driving tests in the car and were pleasantly surprised by its flexibility and the ease with which one could undertake the tests, without any discernible difference. It is important to remember that you do not get as much retardation from simply backing off the power as you do when in a lower gear in a manual car. Even on cross-country routes the car is stimulating and fun to drive, although it comes into its own if you do a great deal of town and long distance motorway driving. Finally, as a personal choice the addition of ABS brakes would be a premium selection.

SUMMARY

The choice of an automatic or manual gearbox for your Porsche is not one which should be governed by fashion or trend. It is almost entirely a choice to be made on the grounds of the type of motoring you will do, and your own preference. With the modern automatic gearboxes currently available on the Porsche 944 and 928S series 4, the discernible differences in road performance are very marginal. The advantages to be gained from the effective use of an automatic gearbox, in the manner we describe above and in Chapter 1 will, for many expert drivers, be overwhelming.

CHAPTER 11

Porsche Development and Production – The Achievement of Excellence

As one of the world's smallest automobile manufacturers, but the manufacturer of a range of the world's most desired high performance road cars, Porsche has always pursued excellence and exclusivity. Superb build quality, technological excellence, coupled with the most advanced engineering design, are not the product of mass production. They require meticulous attention to detail, the continuous application of research and development, and total dedication. In its 40 years of production history, Porsche has produced almost 700,000 cars and yet its achievements in racing, technical innovation, and automotive research are second to none. Such achievements do not, however, come cheaply, or easily.

The basis for all of Porsche's emphasis on innovation in design stems, perhaps, from the fact that any new car has to have a product life of at least twelve years; constant model changes on a current annual production base of circa 40,000 units are impossible and undesirable, to say the least! Therefore, the original concept must be right and constant developments have to maintain the product's lead in technology and performance. The modern high performance road car has become increasingly complex and new technologies such as materials, electronics and aerodynamics are all generating increasing demands upon the automobile manufacturer who strives for performance, build quality, and market leadership in an increasingly competitive arena.

Although some high volume manufacturers have larger research and development facilities, in reputation the R & D Centre of Dr. Ing. h. c. F. Porsche AG at Weissach has few peers. The Centre currently employs nearly 26% of the company's workforce, over 2200 people, of whom over 1000 are engineers. More than 100 engineers and scientists are engaged purely on research, on projects such as engine harmonics, materials technology, and all aspects of automotive safety and performance. Started in the early 1960s, the R & D Centre at Weissach began as a series of test tracks, and by 1974 a complete Porsche factory for Research and Development had been established, and the entire research operation moved from Zuffenhausen. That was, however, only the beginning. Throughout the late 1970s and early 1980s the facilities have been expanded, the expansion domin-

The Porsche R & D Centre at Weissach.

ated by the needs of the Porsche car and its development.

This show place of automotive engineering cannot be supported by revenue from the production of Porsche cars alone, although it is the heart of Porsche's performance and engineering excellence. The Centre's primary task is to support the continuous development of the Porsche product, "Driving in its Purest Form", but the enormous investment in plant and buildings has meant that external research contracts have been essential. However, the company's philosophy of total confidentiality for each client, with the client's acceptance that no engineer can be expected to unlearn research knowledge gained, have provided additional challenges and motivation for the creative and fertile minds of the extraordinarily talented engineers. They also ensure that there is a continuous and invaluable spin-off, from a high level of research and development that could not otherwise be sustained.

The revenue from outside contracts varies but about 30% of the engineering research and development activity at Weissach is on behalf of external clients. Nevertheless this is an essential part of the company's activity, for both technical and financial reasons. The R & D Centre at Weissach has an outstanding reputation and, apart from having undertaken work for virtually every single automotive manufacturer at some time or other, its customers have included the European Airbus Consortium, NATO, and many governments including that of the Federal Republic of Germany.

Statistics can be overwhelming but the following serve to emphasise the commitment that Porsche has to the technological excellence, performance and safety of its products: your car. More than 10% of annual costs are spent on research and development, with more than 30% being supported through outside contracts; DM415 million out of a total turnover of DM3,408 million in 1986/7. Major investments in research and production facilities during the 1980s have exceeded DM1100 million including, at Weissach, a full scale and 1:4 wind tunnel, the most advanced environmental test centre in Europe, a crash facility with the latest recording and monitoring technology (booked by outside clients up to two years ahead), and a new race and test centre for building prototype engines and chassis parts. These are only some of the research facilities that contribute towards the performance and safety of your car.

Porsche build quality is a legend, even amongst hard-bitten motoring journalists, and road test reports consistently praise the way the product is put together. This also requires investment, and new plants at Zuffenhausen in the 1980s include a computer operated parts warehouse, the world's most environmentally favourable paint shop, and a new body plant completed in January, 1988. Hardly surprising that investment has totalled in excess of DM1100 million in less than ten years!

The core of Porsche's activities at Weissach are directed at you, the driver of a Porsche. Their mission is the development of the product and its performance, safety and technological exclusivity. The company's roots are in research and development, and Weissach is building on that tradition. But how does this priority for research and development benefit you, the driver; what does the mission mean in terms of performance, safety and technological excellence? Let us take a few examples.

The safety legislation that many governments have developed has undoubtedly stimulated many manufacturers to design safety features into their vehicles. It is certainly no accident that when an older Porsche 911 from the late sixties was subjected to a modern crash test, it passed. Porsche's new Centre incorporates all the latest monitoring devices and can photograph crashes as they happen, at up to 1000 frames/second. Even the models used to simulate occupants are temperature controlled, and are individually monitored to determine the level of disturbance on impact. In order to ensure constant, repeatable tests, the dummies are standardised into several different sizes, each representing a particular height and weight ratio. Naturally, crashes have to be at precisely controlled speeds and this is achieved by towing the vehicle using a hydrostatic, short-term drive, as the power unit.

A crash impact into a solid object at 110km/h will only last about one tenth of a second. During that brief but expensive period, up to ten cameras will record every aspect of the impact, upon the car and its occupants. The actual recording of the crash impact is another area of high technology, and Porsche has applied the latest techniques to show what happens. An ordinary 16mm camera transports the

Crash testing. Safety comes first.

film past the lens by means of mechanical claws, limiting the number of frames per second that it can take to about 500. By using rotating prism cameras that transport the film past the lens and the prism in a continuous steady flow, speeds of up to 3000 frames/second can be achieved, increasing the amount and detail of crash information. Obviously at such high effective shutter speeds intense light is required. Halogen lamps enable filming to take place both inside and outside vehicles under tests. No mean feat when crashing at such high 'g' forces.

These facilities have enabled Porsche engineers to develop the active and passive safety features incorporated into your car to a very high level of sophistication. Detailed knowledge of vehicle behaviour in a crash identifies not only the damage to actual crash zones, safety cell members, and doors, but also the effect of the impact upon other components. Furthermore, recording the effect on the occupant models, provides detailed information on the effectiveness of internal energy absorbing material, seat mountings, the effectiveness of safety belts, air bags and other restraining devices. The air bag system will probably lead to some redesign of the 911 cockpit interior, as it already has on the 944 for certain markets.

The fact that an early 911 passed current USA requirements is an outstanding example of the safety engineered into every model, but Porsche is never satisfied. As a result of further testing, the safety cage of the Porsche 911 Carrera and Turbo Series cars, always exceptionally rigid, has been further strengthened. Both the roof and door design have been improved by experiments with side-on collisions. Not only will the doors remain closed in both roll-over or side impact crashes at 30 mph, they will also open easily afterwards. The plethora of national legislation, both in Europe and the rest of the world, that governs the specifications of safety, emissions and other aspects of automobile manufacture, rightly requires constant monitoring and testing. Manufacturers now have to ensure that their products comply with this increasingly complex legislation. At Porsche, compliance is not enough, its cars must always be ahead.

The front-engined Transaxle cars have an additional safety feature, in the rigid Transaxle driveline. The combination of front and rear deformation zones within the body structure, allows the rigid axle to transmit impact forces front to rear (or rear to front), outside the safety cell. This is, of course, in addition to all the other extensive safety features on the cars.

Every detail of safety is constantly re-examined, even when only minor changes are made to a car's specification, and thoroughly investigated and proven in hundreds of tests carried out on each model. The company destroys up to 90 cars per year in all of the different safety tests. The advantages of the new crash laboratory at Weissach attract many other manufacturers to use the facility, and Porsche can conduct all the necessary tests to continue the advance of vehicle safety.

The environmental impact of automobiles is another area where Porsche technology is in the forefront of development – in spite of the conflicting demands for power and a low level of environmental pollution. All of the Transaxle Porsche models now have the same power output whether or not they are fitted with an exhaust gas catalytic converter, and they will run on unleaded fuel. In 1971 about 60 emission tests were carried out on cars for the American market but the number has now risen to over 4,000. The new MZU Environmental Centre at Weissach was opened in 1982, as a completely enclosed environmental test centre with six main test and monitoring stands. Previously only about fifty personnel were employed in this area but when the new centre opened over two hundred engineers and technicians were required. This centre played a key role in the development of the three-way system to overcome emission control problems and power loss when an exhaust gas catalytic converter was fitted. The importance of their work is not to be underestimated, not only because regulations that cannot be met restrict markets but because Porsche recognises the need for responsibility towards our common environment. Moreover, Porsche's research has enabled it to meet the demands of both its customers', and the environmentalists.

Some of the most dramatic advances in engine management have been as a result of the application of electronic technology to engine management systems. Without these systems, it is doubtful that such enormous progress would have been made in engine efficiency, increasing the power output whilst not increasing fuel consumption or the level of emissions. The engineers responsible for running the environmental test centre have also been involved in developing the applications of this new technology. Using a microchip known as E-Prom, or Erasible Programmable Read Only Memory, it is possible to monitor and control, precisely, a total of over 4000 items of information, covering both the ignition and the fuel injection system and oxygen sensor (if fitted). At the MZU Environmental Centre's test stands cars undergo tests in controlled conditions which can simulate any combination of climate, altitude and humidity levels. Using the E-Prom chip an almost infinite combination of fuel, ignition, timing, turbocharger boost settings can be constantly monitored, and the engine run at optimum levels to satisfy either emission criteria, power criteria, or a combination of both. The flexibility of the system is extensive and impressive. By exposing the chip to ultraviolet light, its previous programmes are erased. Then, using an electronic keyboard, the chip is re-programmed. A further advance is an EE-Prom which stands for Electronically Erasable Programmable chip! This can be reprogrammed, simply by using electronic means.

The electronics are, perhaps, unimportant details to you the driver, but in practice they are vital links in a chain of command that ensure the Motronic engine management system, controlling ignition and fuel injection, works at peak efficiency; ensuring optimum engine performance. The system is primarily responsible for Porsche's ability to overcome the problems associated with catalytic converters, although the decision to use four-valve per cylinder engines was a major factor in improving power output. It is interesting to note that in America in 1986 Al Holbert drove a 928S series 4, fitted with a catalytic converter, at over 171 mph on the Bonneville Salt Flats to establish a new speed record. In the 1988 model year just over 85% of Porsche's production was delivered with converter technology, the majority without power disadvantage.

The use of new materials, ceramics, plastics, as well as special metal treatments, has also been pioneered at Weissach. The development of the Longlife car in 1973, led directly to the use of special hot dip galvanised sheet steel on all of Porsche's range, and to the current added benefit of a ten-year warranty against rust penetration for all models. Porsche's thinking on the life of the automobile suggests a life of perhaps twenty years or a minimum of one hundred and eighty thousand miles; new types of materials will certainly play their part in such a future. Ceramic port liners have been used in the 944 Turbo to further enhance exhaust emissions.

Another area where Weissach's research and development expertise has led directly to improvements in road car performance is in engine development. Perhaps the most dramatic improvements in this area have been in the use of four-valve engine technology. One strong reason for choosing that route was the decision in 1983 that all Porsche road cars would offer identical engine performance, regardless of catalytic converters; another was the combination of power and torque improvements available. Porsche also recognised that although only the United States of America, Australia and Japan had passed legislation, other markets would follow and Germany, in particular, was taking the lead in Europe. The move to four valves per cylinder will probably extend to other models in the future. The advantages of a central plug position, more available valve area and better fuel burn with lower octane fuel, are very positive.

An example of Weissach involvement in power train development, and one currently receiving considerable attention, is the new PDK transmission. Although the Sportomatic semi-automatic gearbox was not a resounding success on the 911 cars, the new system appears to have all of the advantages and none of the disadvantages of the Sportomatic system. The new PDK transmission (PDK stands for Porsche Doppel-KupplungsGetriebe) fitted to a Porsche 962C race car consists of two shafts each with its own clutch, each carrying three gears. The gear selection system consists of an electrical switch operated by a gear lever. Pushed to the right it is in neutral with both shafts disengaged, pushed to the left and back it selects first. Moving the lever forward selects taller gears sequentially, moving the lever back selects lower gears in order. Two small buttons on the steering wheel will replicate movements of the lever; top button for higher gears, lower button for lower gears. The gear selected and the gear actually engaged is indicated on a digital display directly in front of the driver, e.g. the gear engaged is first, that selected second. Thus the system selects the correct shift point and allows gear changes to take place without lifting off the throttle

The system has already been used with considerable success in both the World Sports-Prototype Championship and West German Supercup series fitted to a Porsche 962C and has been developed over a four-year programme. The thought of a return of the semi-automatic transmission may raise the eyebrows of some purists, but it certainly appears to be an excellent driving tool in the hands of Hans Stuck who managed to win the West German Supercup series in 1987. This development will require further refinements in the future before it can be considered for incorporation into a road car. With both electronic and mechanical engineering playing their part, it is yet another example of technological combination, which is an essential ingredient of modern technogical progress.

The cross-fertilisation of ideas, and continuous combination of dif-

The Porsche 962C with PDK semi-automatic gearbox.

ferent technologies, is an increasingly important feature of modern automobile engineering. Advances in new materials, electronics and of course, aerodynamics have all played their part in modern motorcar design. Furthermore, where they are not used on the car such technologies are often applied in other ways, such as monitoring performance, calculating stresses, or identifying a preferred combination of a number of variable factors, as in fuel management systems. Examples of this cross-fertilisation abound at Weissach, but human intellect has to weld it all together. The new wind tunnel facility and the environmental test centre are both working closely on the problems of fuel consumption, together with, of course, the materials laboratory and engineering.

Porsche is continually improving the car's performance but it is concerned that the overall concept of the whole car is improved, not just speed or power. Outside engineering contracts are never pursued, they wait for the clients to come. The Porsche high performance road car range and its competition activities are its raison d'être and the bearers of the Porsche message of engineering excellence around the world.

A primary example of this pursuit of total performance improvement can be seen in the results of wind tunnel tests, previously undertaken at Stuttgart University, but now carried out in Porsche's new full size wind tunnel at Weissach, built at a cost of DM37 million in 1986. Since aerodynamics is, along with weight, engine design and management, a fundamental element in improved performance

without sacrificing fuel consumption or emission controls, the tunnel is another powerful element in the determined search for perfection. The full size tunnel is now used alongside the scaled down tunnel which can take one quarter scale models. The whole facility incorporates, of course, the latest monitoring equipment as used in the aeronautical industry. An example of the value of the wind tunnel's contribution is the difference between Porsche's latest jewel, the 959, and a current 911. The Cd value of the former is 0.32 against 0.385 for the 911. One of the many factors that contribute to the astonishingly low fuel consumption of the 450 bhp supercar. More importantly still for the 959 is that it has zero lift for maximising performance handling.

The increasingly intense search for low emissions, without sacrificing power, pursued by Porsche so successfully within the last five years has led to the incorporation of several developments that have had the effect of reducing fuel consumption as well as the level of emission. However, the search then began for lighter materials, and better body designs that will reduce drag, and therefore the power required for a given speed. An example of the progress made through this combination of different technologies is in the development of the 911 and 944 Series. Launched in 1963 and 1981 respectively, both model types display how the pace of improvements have quickened.

In the twenty five years between the 2.0-litre 911 of 1963 and the current 3.2-litre 911 Carrera, the acceleration time from 0-60 mph has fallen from 8.7 seconds to 6.1 seconds, maximum speed risen from 130 mph to 152 mph, and yet fuel consumption has improved. Similarly, the 944, launched in 1981, then produced 163 bhp without catalytic converter, and had a 0-60 acceleration time of 8.5 seconds. The 944S, launched in 1986 produces 190 bhp with or without catalytic converter, has a 0-60 acceleration time of 7.9 seconds, and a fuel consumption virtually equal to the original 944. There is little doubt that such progress will be maintained and improved, and indeed it has, since the 450 bhp Porsche 959 has a fuel consumption almost equal to the 911 Turbo!

However, all of this activity and research is, of course, undertaken with driving in mind. Practical driving tests must be the final arbiter of any technical advances, and the proving of engineering innovation has to be undertaken under road conditions. Stationary test facilities are only capable of providing a parallel control; developments have to actually work in practice, on the road or track. Your car, or in fact one like yours, has been the subject of a large number of driving tests, which is one of the ways in which both product quality and innovation are examined, tested, and developed under practical conditions.

At Weissach there is a team of over 40 engineer drivers who undertake all of the tests which are used to evaluate innovations,

Opposite The new full-scale wind tunnel makes a vital contribution to aerodynamic design.

or to monitor the performance and product quality of current production models. However, their influence does not, by any means, cease at that stage, nor does it start there. This is perhaps the last, but by no means the least, of the exceptional research facilities at Weissach. The Porsche philosophy has always been dominated by the concept of a car for drivers, never simple utilitarianism, pure speed, or simply comfort. Always the aim has been that elusive combination of performance, handling, flexibility and aesthetic appeal in a final product that is, above all, a practical, truly road-going car that can:

> "compete in the East African Safari or at Le Mans, drive to the theatre and finally in New York city traffic",

to repeat the words of Professor Dr. Ferry Porsche.

The secret, or rather one of them, is the influence that engineer drivers, or enthusiastic drivers who are also automotive engineers, have had on the company's products, from concept through design and development, production, and then the continuous/on-going development programme. It is significant that the firm's founder could well have followed a career as a racing driver, and that all of the company's leaders speak eloquently about cars, in a manner that confirms their enthusiasm, not just for selling them, but for driving them.

The concept of each of the Porsche models has been determined by the ideal of these engineer drivers, people who have the ambition to produce cars that they would want to own and drive. The team of drivers responsible for testing the results of their colleagues' ideas, for monitoring the quality of the product from the production lines, and for working on concepts, are all engineers; as one of the Chief Test Pilots (their title; as in Italy, drivers are pilots) said "we are ultimately responsible for judging the behaviour of the car". Their response and feedback is crucial to the handling characteristics and performance of each car. There are no armchair designers at Porsche. The principles that determine the desirable characteristics of a high performance road car have, of course, changed. They are a function of the available technology, and the combinations within which they can be incorporated into a final design. The early cars were designed within the concepts of the times, and in the 1950s, high performance road cars were designed with inherently oversteering characteristics. Thinking has changed and Porsche now regards the desirable characteristics to be 'early' understeering cars, with characteristics as neutral as possible under increased cornering forces, and then a predictable and gradual transition into oversteer at extreme cornering forces. The latter should not be experienced under normal road driving conditions.

The Test Department undertakes three basic types of testing. Firstly the constant test programme of all current models, selected at random from the production line. Secondly, a programme for modifications on existing models, including such features as any changes in original equipment, particularly specifications for tyres. Thirdly, the department is constantly engaged in prototype testing, both in a series of special vehicles, and in new designs. We will deal with each programme in broad outline.

The production test programme is governed by a continuous search for highest build quality and consistency. The series of tests starts when the car is pulled out of the production line, and is a repeat of all the final production checks that are carried out. Every detailed deviation (if any) is recorded and will be passed back to the factory. The driving tests are a series of predetermined examinations of the car's performance and handling which take place on the extensive test tracks and skid pads at Weissach, particularly the handling circles. The basic handling tests are meticulously recorded, and indeed photographed. Each test is designed to demonstrate a particular feature of the car and the results are compared against standards that are determined at the prototype stage. The handling tests consist of a large number of exercises on the 40, 60, and 190-metre skid pad circles at Weissach. Specification, speed, road conditions and all up weight of the vehicle are carefully recorded. Therefore, a detailed pattern of slip angles, lateral acceleration 'g' forces, braking 'g' forces, roll angles, and steering forces required can be measured. At each stage the results are examined against the predetermined norms for

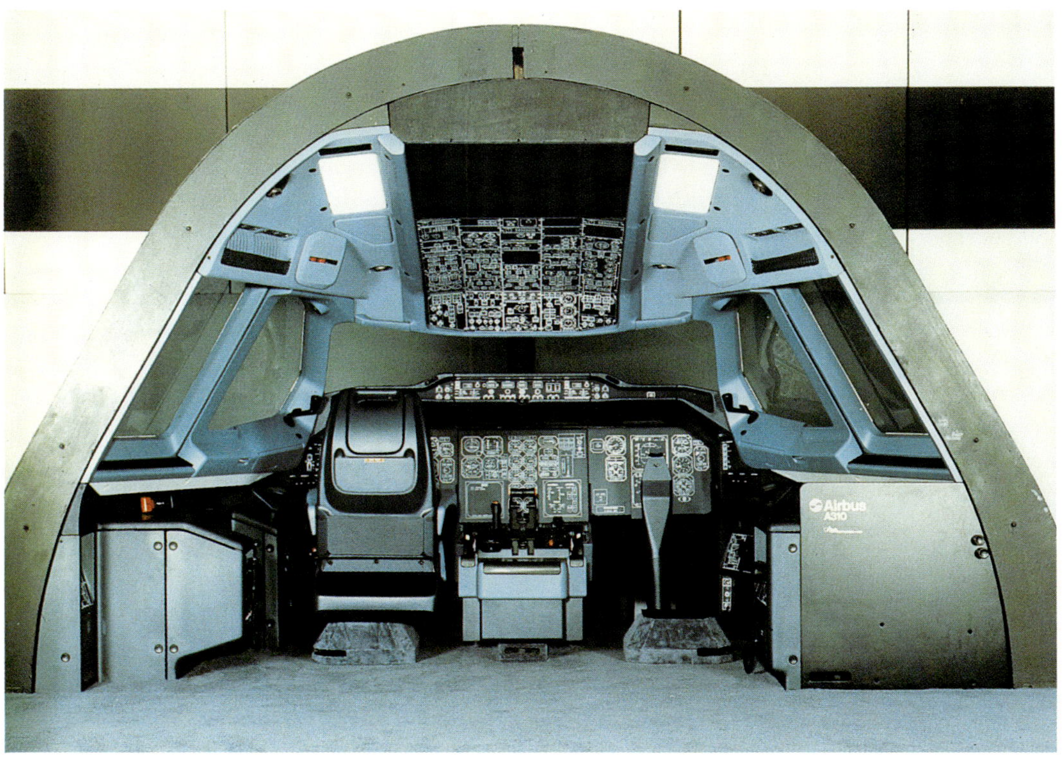

The A310 Airbus cockpit. Research and development work is not confined to just land, but also sea and air-based vehicles as well!

the particular model. These constant speed, constant radius tests will rapidly identify any abnormalities in the car's suspension assembly. They also ensure that each model has a very high consistency of handling; there are no cars other than excellent ones.

The second stage of production testing takes place on the pavé and 'mountain' track at Weissach, and consists of more pre-determined tests designed to identify any faults in production. These are not destruction tests but are designed to determine that build quality is being maintained. The important common factor is that the same detailed model specification and weather conditions can all be repeated; vital for a common base for comparison. This standard programme includes performance, acceleration, braking, handbrake, and fuel consumption tests in order to compare similar specifications, against pre-determined standards.

Finally, a number of other tests are carried out at random in different climates. The performance tests include taking lap times under common conditions, and checking the % speedometer/rev. counter error. Porsche's tolerance is very low compared with other manufacturers; which is why the manufacturer's performance figures are always pessimistic! These augment the climatic testing already undertaken during the model's development stage. Naturally, throughout these tests, the reports are completed by the engineer, or engineers, who perform them.

The long range tests carried out require a car to be driven on normal roads and other tracks, all over the world, but primarily in Europe. The 16,000 kms at Weissach are supplemented by 80,000 kms elsewhere which includes over 12,000 kms of town driving, 20,000 kms of country driving, 24,000 kms of autobahn driving, 6,000 kms at the old Nürburgring circuit plus 18,000 kms at the high speed track at Nardo in Italy. At every stage detailed measurements are taken of all the major and minor factors concerned with performance.

Only by such long-range testing can, for example, wear rates on engine, chassis, suspension, interior trim, external paintwork, tyres and wiper blades, be assessed on the road tests. Laboratory tests are, in fact, undertaken to do all of this, including a computer based suspension testing facility, but Porsche insists that drivers are the ultimate arbiters of what is good, and bad.

The important factor governing all of these tests is consistency and the frequency with which the cars are selected. Consistency is ensured, not only by the comprehensive facilities available at Weissach but also by the exceptionally detailed recording system which the engineer drivers have developed, a system probably unrivalled by any other manufacturer. Additionally, Porsche tests a far higher proportion of its production series than most manufacturers.

The second series of tests covers any developments in engine, drive-train, suspension, braking, or chassis design. With the massive amount of data available from the previous series of tests, any single change can be assessed and judged against very clear criteria. Changes for the sake of fashion are eschewed by drivers who are searching for driving pleasure. The amount and consistency of all the information is enormous and is, of course, now computerised. These tests are very similar to the production driving tests, with the exception that the basis for comparison is, inevitably, more complex. Isolating what characteristics are affected by a particular innovation can be a problem and will often expose areas of the car where improvements are needed to cope with new developments in particular areas. Whilst laboratory tests will isolate specific improvements, it is only under realistic road conditions that components can be evaluated within the complex system called the automobile.

The final testing programme is rather less routine than the previous two, although ultimately any car will be submitted to the same battery of tests, and a great many more. Porsche has developed a special vehicle which, in addition to being able to test specific new components, can be so radically altered as to be capable of testing different engine, suspension, drivetrain or bodywork layouts. Nicknamed, rather unglamorously, the test mule, this vehicle has replaced the practice of using individual production vehicles in which to slot new components for testing. The official name for the vehicle is Porsche Experimental Prototype, inevitably shortened to PEP.

As technology has progressed, the design of such a vehicle became essential since production vehicles themselves have limitations; it is difficult to test a front-mounted engine in a 911 bodyshell. The PEP vehicle has a modular (it comes apart in bits!) design, which allows the engineers to alter the basic characteristics of this test rig very simply. Although, from photographs, it has been mistaken as a new prototype, don't be fooled, it is not. Front and rear axle configurations can be adapted and, due to a telescopic Transaxle tube, even the wheelbase can be altered. Any of the basic control functions including steering, braking and throttle control can be monitored, and the external aerodynamics of the vehicle altered by changing the body panels. It is even possible to alter chassis rigidity, a crucial factor in assessing handling.

When first conceived, this vehicle was set-up in order to simulate a number of existing models for which detailed test data was available. By using the records of test results, described earlier, it was then possible to assess the accuracy of the vehicle as a simulator. Needless to say the comparisons were favourable, and the PEP vehicle is now a well used and very valuable element in the test driving programme.

Our visit to Weissach was an opportunity to experience the atmosphere of this extraordinary temple to automobile development, and one cannot fail to be impressed. The essentially practical and business-like atmosphere of the place is obvious, as is the total enthusiasm for the product of all who work there. Indeed the total organisation

Porsche Experimental Prototype vehicle.

of Weissach is, in our view, an object lesson in the use of talent to pursue a clear objective – engineering excellence. The nine departments all report to the Director, Professor Dr. Helmuth Bott. The organisation's canteens and restaurants are all in one building, the hexagonal Gästecasino, and departmental contacts are continuous. Work is not set in closely defined departments, but is often referred to a number of departments; communication is a Porsche watchword.

Weissach offers tremendous advantages for you, the Porsche driver; advantages that are derived from the superb facilities, the huge concentration of engineering talent, and the stimulus of external links the Centre maintains with universities, and other engineering research organisations. The company's early history as a research and design consultancy means that the philosophy of research in engineering has always been driven by the requirement for client satisfaction. At Porsche, engineering may well be considered an art form but it also has to satisfy the demands of the client, the end user, you the driver.

The needs of state-of-the-art technology which are included in the modern high performance motorcar, and in Porsches in particular, demand intense cooperation between all of the technical and engineering disciplines. The central location means that the nine different departments which are powertrains, chassis development, styling, research/advanced development, competitions, outside contracts, military development, company development, and Porsche passenger car development are all able to draw on their joint experience. Furthermore, the separation of the main automotive research departments from those that are responsible for external contracts, military development, and Porsche passenger car developments, enables a unique cross fertilisation of concepts, ideas, and solutions to take place. Under the inspired direction of Professor Dr. Helmuth Bott, the heads of the nine departments, all 'Porsche people', are able to pursue the goal of "driving in its purest form".

CHAPTER 12

PORSCHE RACING – A TRADITION OF DEVELOPMENT

Racing is a Porsche tradition since, from the beginning, Professor Dr. Ferdinand Porsche (as he was to become) regarded involvement in motor sport as an important part of the development process. Ferdinand Porsche's first racing car was based upon his designs for the Lohner-Porsche electric car, and on the 23rd September, 1900, he established a 10 km record at over 40 kilometres per hour, up a gradient of 1 in 25! During the 1920s and 1930s, when motorcars were increasing in popularity, it was common for manufacturers to be involved in racing; racing did improve the breed, even Grand Prix racing. Ferdinand Porsche played an important part in this tradition. In 1910, as designer, mechanic, and winning driver, he was responsible for the Austro-Daimler 1-2-3 win in the Prince Heinrich Trial for international touring cars. Ferdinand Porsche was then just 35 years old and already Technical Director of the Austro-Daimler company.

In succeeding years, Ferdinand Porsche's continued involvement with racing was inevitable, distinguished by success and the application of many racing innovations to high performance road cars. The lightweight Sascha, a 4-cylinder, in-line engined sports racing car designed by him, contested the Targa Florio in 1922 and surprised many people by coming seventh in the large sports car class with a bored out 1100cc engine. The tradition of lightweight cars with excellent handling characteristics was already established.

Then, in 1932 Porsche decided to design the now legendary Auto-Union Grand Prix racing car to the new racing formula ratified by the International Sports Commission in Paris. This unlimited formula had a weight limit of 750kg and unlimited engine capacity, a formula which Porsche approved of as he felt it gave the designer maximum flexibility and freedom. The layout of this car with its rear engine, rear-wheel drive and central steering was to have a profound influence on high performance road car design as well as Grand Prix and sports racing car design throughout the rest of this century. Look at any modern Grand Prix car and you will find that it is based on this design layout.

Porsche's view that the activity of racing provides an efficient testing ground for new designs and concepts was and is undoubtedly

Bernd Rosemeyer in the Porsche designed Auto-Union Grand Prix racing car.

sound. Indeed such was the strength of his view, and his enthusiasm for motor racing, that Professor Dr. Ferdinand Porsche pursued the idea of a sporting version of the VW Beetle whilst he was designing the car in 1934 and 1935. This sporting version was, in fact, designed by Porsche immediately after the first VW Beetle Series 3 cars were actually running in 1937. Unfortunately, it was repeatedly turned down by the German authorities because the VW was to be seen as a low-cost, people's car. However, a 1500cc engine for the car was already more than just a vision in the mind of this extraordinary man. The authority's continued refusal did not prevent Ferdinand Porsche from pursuing his idea and three very similar designs were prepared numbered 114, 114K1 and 114K2; they were shelved in September, 1938. The proposed Berlin-Rome rally changed official policy, the go-ahead was given and three cars were built based upon these designs, which bore many similarities to the first Porsche. The rally did not take place but the cars remained at the Porsche works, then established at Zuffenhausen; the last of them still exists.

Porsche's motor racing tradition was, as with many other things, interrupted by the Second World War but it was a commission to design a new Grand Prix car that helped establish Porsche as a motorcar manufacturer in its own right. The famous Italian company Cisitalia, commisioned four designs in December, 1946, which included a Grand Prix car and a 1500cc sports racing car; the former was considered the most important. This quite remarkable advanced design had a horizontally opposed, 4-camshaft, 12-cylinder engine, rear or four-wheel drive (engaged or disengaged by a lever in the cockpit) and three superchargers. This advanced design, the brain-child of 'Ferry' Porsche, had its engine, clutch and five-speed synchromesh gearbox in front of the rear axle. A truly astounding number of advanced features on a single, new design. Unfortunately, Europe was not to see this car race in post-war Grand Prix, as Cisitalia went bankrupt, but all of the innovations have subsequently appeared on high performance road cars.

One month after the first 356 was completed, it won a race at Innsbruck in 1948 driven by one Herbert Kaes. The car was chassis number 356-001 and the engine number was 356-2-034969. Porsche's manufacturing and racing tradition had begun. In 1950, Prince Joachim zu Furstenaberg and Count Konstantia Bercklein entered an 1100cc Porsche 356 in the Swedish Midnight Sun Rally and won the event outright. During the same year another owner, Otto Mathé, a one armed driver from Austria, raced and rallied an 1100cc coupé. A large number of Porsche owners have always raced their cars, with very considerable success.

The next racing landmark for Porsche was to prove, eventually, the basis of its sports racing car success; Le Mans. Unfortunately, Professor Dr. Ferdinand Porsche did not live to see the first production car carry his name in a major international event as he died on 30th January, 1951. Three 356s were entered in the 1951 Le Mans 24 hours at Sarthe. The cars were all fitted with a special camshaft designed by Dr Ernst Führmann (later to become Chief Executive of Porsche from 1972 to 1980), which increased power from 40 to 48 bhp, and aluminium bodies and spats over the wheels, to improve streamlining. Only one finished after practice accidents to the other two. However, the Porsche 356 driven by Veuillet and Mouche won its class and came fifth in the Index of Performance.

The Le Mans class victory of 1951 was, according to Professor Dr. Ferry Porsche, an object lesson. Firstly, if possible, always race with more than one car. Secondly, any superfluous weight is a handicap. The latter lesson has since been applied to all Porsche race and road cars, without sacrificing safety or comfort. High speed, reliability, exceptional handling and acceleration, all achieved with a comparatively modest engine size and lightweight chassis has become a trademark of Porsche high performance road cars.

The year of its first participation at Le Mans the company was producing just 1169 cars. Shortly after the race, Professor Dr. Ferry Porsche appears to have taken the view that racing was good for advertising the company's products and provided the opportunity to test cars more efficiently than on the factory bench. Certainly, the savings from advertising could then be spent on both racing and testing. In 1951, racing was less expensive and there was still a direct

Porsche's first Le Mans success, a class win in 1951.

relationship between the cars on the track and production models. However, both philosophies have been continued at Porsche and it is certainly true that road and track vehicle testing is used to prove cars.

Porsche racing success continued and in 1952 two Porsche 356s won their respective classes in the famous Mille Miglia road race. The 1952 Liège-Rome-Liège Rally was dominated by Porsche 356s which took the first, third and fourth places, and a 356 finished eighth in the Carrera Panamericana at an average speed of 83 mph. These first Porsches were always popular with private entrants. As Ferry Porsche said:

> "Our starting goal was to build a sports car of modest dimensions which could cover long distances at high speeds without tiring the driver or passenger".

In 1949 a VW dealer in Frankfurt had built an open racing two-seater, based upon Porsche 356 components. This car was known as the Glöckler Porsche and was the first Porsche to carry, as a prefix, the name of its private entrant and builder. A tradition which has resulted in many major race successes and continues today. The Glöckler-Porsche won the German Drivers' Championship in 1950, and the following year won the same championship in the 1500cc class. However, in 1952 the competition became too much for a private entrant to cope with, which is why work began on the first Porsche racing coupé and Dr. Ernst Führmann started on the new four-camshaft 1500cc engine.

In 1953, two of the new cars were entered at Le Mans, both fitted with an enclosed, streamlined, body shell. They crossed the finishing line together and von Frankenberg won the 1500cc class. During practice for the race a Spyder, powered by the new 4-camshaft engine was tested, the first time this engine was so used.

During 1951 and 1952, Porsche designed the now famous baulk-ring synchromesh gearbox, and it was introduced into production cars in the latter year. A spin-off from its involvement in the 1953 Le Mans was that the company witnessed the use of Dunlop disc brakes. As Professor Dr. Ferry Porsche recorded, "We sat up and took notice". Primarily concerned with reliability and consistency, Porsche fitted disc brakes to its first production 356 Carreras in 1955. From its launch, the 911, and all subsequent Porsche models, have had disc brakes.

The Porsche 550/1500 RS, with the Führmann-designed Type 547 flat four-cylinder, four-camshaft, 110 bhp, 1498cc engine, was first raced in the 1954 Mille Miglia. Driven by Hans Herrmann and Herbert Linge, it proved very reliable, finishing in sixth place. By the end of 1954, the factory was producing 550 Spyders for sale to customers, a Porsche habit which has bestowed a number of benefits over the last forty years. Private entrants joined what has, quite rightly, become known as the Porsche family, and were often given extensive assistance from the factory. In return, the company was able to gain knowledge of its cars' performances, on a much wider basis, than if it had retained designs exclusively for factory use. Nevertheless, this policy has often led to private entrants giving the factory entered cars some surprises, without reducing the company's enthusiasm for private entrants, or its willingness to assist them.

At the Avus circuit in 1954, standard (a slight misnomer perhaps) 550 Spyders were lapping very nearly as fast as current Formula 2 racing cars. Richard von Frankenberg, the Porsche works driver, was lapping the circuit at 123.5 mph, with four cylinders and 110bhp. It is interesting to note that the 1954 Avus Grand Prix was won by Juan Manuel Fangio, in a Mercedes Benz W196, fitted with a normal Grand Prix body, at a speed of 133.5 mph. This car had an 8-cylinder engine and produced over 260bhp! Porsche was, and still is, extracting phenomenal performance out of lightweight cars, with modest power output, that handled exceptionally well.

The notable success of 1957 was fifth place in the last of the great Mille Miglia road races, with a 1500 RS. In the 1958 season, Porsche concentrated its efforts on running the modified 1500 RS-K Spyder, which was, of course, available from the factory for private buyers. Successful in both road and track races, they collected notable third, fourth, fifth and ninth places in the 1958 Le Mans. Only a 3-litre Ferrari, and an Aston Martin, could beat the small 1.5-litre cars. Porsche was to wait three years for a comparable result.

In 1958, Porsche ran a 1500 RS-K with central steering, for the first time, in the Formula 2 race at Rheims. Apart from central steer-

Porsche 550 Spyder at the Nürburgring driven by Richard von Frankenberg.

ing, this was a perfectly standard Spyder and driven by Jean Behra it won convincingly. In an attempt to repeat this success a year later, Porsche took two racing Spyders with exposed wheels. Edgar Barth finished third, behind Moss and Hans Herrmann in a Behra-Porsche. In the same year, Porsche was second to Ferrari in the World Sports Car Championship, further confirmation that the marque was a force to be reckoned with. Incidentally the Porsche 911 still uses a much developed steering system which operates from the centre of the shaft, from two universal joints; giving excellent feel and safety.

1959 was not an auspicious year for Porsche in road racing although a Porsche 1500 RS-K won the Targa Florio driven by Barth and Wolfgang Seidel. At Le Mans, both the factory and private entrants retired, all with broken crankshafts. 1960 was marginally better, with the only Porsche to finish at Le Mans a privately entered 1600 GS Abarth Carrera which won its class. In the 1960 season, Porsche reached a crossroads, or the end of a 'Block'; the 4-cylinder, 4-camshaft engine had reached the limit of its development potential. Formula 1 development problems were taking up a great deal of engineering time and its 8-cylinder engine development programme was behind schedule.

Porsche's first official entry into Formula 2 racing was in 1959, when Wolfgang von Trips drove a car in the Monaco Grand Prix; he hit a wall on the second lap and retired. The 1960 season was considerably more successful and Porsche won the Formula 2 Con-

Porsche's 1962 Grand Prix car.

structors' Championship when Jo Bonnier was victorious in the Formula 2 race at the Nürburgring. In April, 1961, its first season in Formula 1, Porsche entered the Belgian Grand Prix at Brussels. Neither of its cars finished, but the race proved to Porsche that the 'old' 4-cylinder type 537/3 engine was still competitive, at least against the British cars; the red Ferraris from Maranello were another thing altogether. Porsche's new chassis, with a longer wheelbase and new front and rear suspension, undoubtedly played a part in extending the competitive life of the 1500cc RS engine. Porsche had chosen the Monaco Grand Prix as the opening race for the 8-cylinder engine but it still was not ready. It ran three 4-cylinder engined cars instead for Bonnier, Herrmann and Gurney. Gurney, who came in fifth, scored Porsche's first placing and points in this, the premier formula. In the French Grand Prix at Rheims, Dan Gurney finished second behind a Ferrari. In spite of not finishing at the German Grand Prix, Dan Gurney finished third in the Drivers' Championship and Porsche came third in the Manufacturers' Championship.

The 1962 season started on a much more positive note, with the 8-cylinder engine set in a completely new Grand Prix chassis which included disc brakes of Porsche design. The new car appeared for the Dutch Grand Prix at Zandvoort. Unfortunately, Dan Gurney took to the sand dunes and although the car proved reliable, he finished a long way down the field.

The French Grand Prix at Rouen was a very different story. Gurney was on the pace in practice and took sixth place on the grid. Driving a very steady race, in a car which ran with true Porsche reliability, he finished ahead of the field. Porsche's first Grand Prix victory was, unfortunately, to be its last for many years, and so far its last as a constructor. At the end of the 1962 season Porsche announced that it would only take part in occasional Formula 1 races. This proved impossible, and the company has not, to date, entered a chassis in Formula l racing since the American Grand Prix in 1962.

1963 was a watershed for Porsche, since not only did it launch its new 911 high performance road car but it also scored its first victory with the new 8-cylinder racing engine, in the 45th Targa Florio. In its final, official, successes the 550 Spyder, the first pure sports racing Porsche, carried Edgar Barth to victory in the European Hill Climb Championships in 1963 and 1964, demonstrating its superb handling and versatility; characteristics inherited by all modern Porsche road cars.

In spite of withdrawing from Formula 1 and 2 racing, Porsche's determination and enthusiasm for sports car racing was never in doubt. The company's whole development philosophy demands that its high performance road cars and its engineering expertise are pitched against the world's best. Competition, as Professor Dr. Ferry Porsche has said, is still a vital element of development; bench testing still has to be assessed against the rigours of practical road tests and track racing, in the hands of driving engineers and enthusiasts. Throughout this period, Porsche's racing and development philosophy had been closely tied to advancing the performance and technology of its high performance road cars and it had always participated in racing with modified production cars, or in the case of the Spyder, with a modified chassis and race developed engine. But a new age was about to begin. In spite of the success of this development philosophy the world was changing. Other manufacturers were pursuing success in sports car racing for reasons of publicity, as well as longer term technological advance.

As an early warning of the change in its racing philosophy, Porsche launched the 904 sports racing car. The Porsche 904 was another effective and beautiful design, from the mind and hand of Ferry Porsche's son, Ferdinand, nicknamed 'Butzi'. Initially, it was intended that the new 6-cylinder boxer engine used in the 911, would power the car, but many early examples had the existing 4-cylinder engine, straight out of the Porsche 550 Spyder. This new car was a break from Porsche tradition, with a tubular box-frame chassis of great stiffness and a fibreglass body. Private owners bought the cars with the Type 547 engine, but the works cars were all fitted with either the new 6-cylinder units, or the 2-litre, 8-cylinder, racing engines. The first twelve cars were sold to private owners in the United States and success soon came, both there and in Europe. A 904 car won its second major race, the Targa Florio, in 1964 with another in second place; shortly after, a class win was recorded in the Sebring 12-hour race in America.

The Porsche 904 was also designed by 'Butzi' Porsche.

Racing cars, particularly sports racing cars, are rarely sufficiently flexible to handle well in rallying and road racing; the 904 was. In January, 1965, a Porsche 904 took second place overall in the Monte Carlo rally, in the snow! In the Targa Florio later that year, all of the 904s powered by the 8-cylinder racing engine retired, but a 904 powered by a 6-cylinder boxer engine developed from the 911 unit took third place. The conclusion drawn from this experience was that the latter unit was more reliable for long distance racing. The car was not really fast enough to win Le Mans, but, in the same year, it still managed to take fourth and fifth places.

Throughout 1964 Porsche was preparing its new high performance road car, the 911, also designed by 'Butzi', for racing. In 1965, two Porsche engineers, Falk and Linge, (Peter Falk was to become the Head of Porsche's Competitions Department at Weissach) took the 911 to a convincing class win in the Monte Carlo rally of that year. The mild, low speed understeer and slightly quick change from under to oversteer, which was only evident at the high lateral 'g' forces this car was capable of, were eliminated as a result of suspension developments through rallying. The car's road and track racing, and rallying results are evidence of its superb reliability, speed and handling.

When Professor Dr. Ferdinand Porsche's nephew, Ferdinand Piëch took control of research and development in the late 1960s, Porsche entered the field with a winner, the 917. The years from 1969 to 1973 were therefore without any doubt a period of Porsche triumph, both on and off the track. The 917, christened 'the ultimate weapon', powered Porsche through to three consecutive World Championship of Makes from 1969-1971, winning 15 of the 24 World Championship races for which it was entered. In 1970, the most coveted prize of all was delivered to Porsche by the 917. Hans Herrmann and Richard Attwood achieved the first outright win for the company at Le Mans. Another Porsche 917 repeated this success in 1971.

The United States had always been an important market for Porsche, with many fans and some very successful private entrants racing there. Porsche started to prepare a car for the North American Can-Am Championships because it considered that a well prepared 917 would be successful. The decision was partly prompted by the FIA's decision to impose a 3-litre limit on cars in World Sports Car racing and by the decision to include Can-Am races in the FIA calendar. Porsche decided to follow its successful experience with John Wyer, who had run a very successful works sponsored team, and made similar arrangements with the Penske Racing organisation in the USA.

The 917 was given a 4.5-litre engine with twin turbochargers, for the 1972 season, in order to try and match the 8-litre power of the all-conquering, until then, McLarens. This turbocharged racing car was a new venture for Porsche since, although it produced 850bhp,

The Porsche 917, christened 'the ultimate weapon'.

and 1100bhp in 5.4-litre form, technology was still at an early stage, especially in wastegate control.

Once again, Porsche turned to practical driving tests to develop the turbocharged unit. The complex relationships between turbocharger and engine were gradually resolved, and the critical response time from accelerator depression to turbocharged input, was reduced to completely unexpected levels. It was as a direct result of these racing developments that Porsche was able to launch the 911 Turbo.

The Can-Am challenge proved highly successful, Porsche wrapping up the title in both 1972 and 1973. Additionally, Mark Donohue in the 1100 bhp 917-30 lapped the Talladega track at 221.112 mph to set a new closed circuit record which lasted for nearly fifteen years!

In 1974, Porsche set in train the next stage of development for its high performance road car, the Porsche 911. The engineers decided to exploit their experience of turbocharging, gained with the turbocharged 917 and soon to be applied to the production 911 Turbo, by using a turbocharged six-cylinder engine in a Carrera RSR. During 1974 their aim was to gain development experience both of the turbocharger application and of the running gear and aerodynamics. This would, they hoped, put the company in a strong position for production based racing, demanded by future Group 5 regulations. They were certainly not disappointed.

In spite of early aerodynamic problems with a car producing 500 bhp from a 2.1-litre, 6-cylinder engine, the Porsche Carrera RSR-Turbo came second at Le Mans, only 53 miles behind the Matra, and second in the Watkins Glen 6-hour race. However, the new Group 5 rules were postponed for a year and so Porsche withdrew from racing for 1975 and the 3-litre Carrera RSRs in private hands carried the Porsche flag. The Porsche 911 Turbo was launched in Paris in October, 1974, and its turbocharged engine was a direct result of the experience gained with turbocharging the 917 Can-Am Championship cars in 1972/3. Furthermore the 911 Turbo relied on 917 derived disc brakes and calipers for its stopping power.

Porsche's philosophy of using racing and rallying as a development platform has always worked extraordinarily well. The improvements made to the 911 Series were often a direct result of racing or rally experience. However, by the late 1970s, the company was influenced by the increasing technical divergence between competition and road cars, that was becoming more extreme. Although many innovations were taking longer to develop for road use, Porsche continued to apply its technological developments from road and track racing experience to its high performance road cars. Aerodynamics and engine emissions were the next major area of improvement, the latter being particularly important because of the more stringent American regulations that were being enforced.

During 1978, the factory developed a 4-valve per cylinder head for the 6-cylinder, 911 production engine, exclusively for the factory

Porsche 935/78 'Moby Dick'.

team 935s. The cylinder heads were water-cooled, although air-cooling was still used for the cylinder barrels. The 2.8-litre engine was the basis for this advance, although a 3.2-litre version was prepared for the 935 of 1978, and a 2.1-litre engine for the 936. With this additional cooling, the temperature of the head was lowered by some 60° Centigrade, allowing higher boost pressures than had been used to date. The water-cooled engines used very few components from the original 911 engines, ran up to 9000 rpm and had a considerable power advantage, with 750bhp at 8200 rpm. This concept was used in the 2.65-litre engines that were prepared for the 1980 Indianapolis race.

At another level, the new generation of front-engined, Transaxle cars was entering competition. In 1979, a 924 (with specially converted suspension) entered Le Mans and won the 2.0-litre class. In 1980, Porsche contested Le Mans with a racing version of the 924 Turbo. These cars were put through an extensive series of tests, including the 1000 km test on the destruction track at Weissach. The three cars performed satisfactorily in the 24-hour event, but the 'racers' amongst Porsche fans and personnel may have thought that the 936 cars would have achieved a better result. The 924 Carrera GT Le Mans cars finished sixth, twelfth and thirteenth but the tests and the race gave invaluable information to the factory on the new front-engined, Transaxle cars.

The position in 1981 was rather different. Porsche had a new Chief Executive at the helm and the world, economically, was looking rather brighter than it had for several years. The ultimate 936 car was produced in 1981 and driven by Jacky Ickx and Derek Bell, it ran like clockwork and won the Le Mans race convincingly. But, perhaps more importantly, Porsche entered a 924 GTP fitted with a prototype 2.5-litre Porsche four-cylinder engine. The engine was essentially half of the 928 unit, with four valves per cylinder and a single turbocharger. This car finished seventh overall and won a trophy for spending the least time in the pits. This development was to reach the road in the guise of the Porsche 944 Turbo.

In 1982, Porsche decided it would build a Group C car to comply with the new thermal efficiency regulations. The challenge was the beginning of another period of Porsche dominance – the latest, but not last, period of technical developments for racing which were rapidly transferred to the road. To meet the fuel consumption restrictions Porsche again collaborated with the Robert Bosch company on fuel injection and engine management systems for the new car. The 956 was Porsche's first full monocoque, which also benefited from intensive wind tunnel research using the full ground effect aerodynamics permissible under the regulations. At Le Mans in 1982, 956s finished 1-2-3 and the car was voted the Motor Sport Car of the Year. The 956 and the 962C dominated all Group C racing between 1982 and 1986, in a manner not achieved even by the legendary 917. Although the factory team withdrew from Le Mans in 1984, the race was won by a private team entrant, Joest Racing, with a Porsche 956C. Between 1982 and 1986 Porsche won the World Sports Car Championship every year. Porsche has a record twelve outright Le Mans victories to its credit since its first participation in the event in 1951; almost 22% of all of the Le Mans races run and more than any other manufacturer. For the statistically minded Porsche's outright wins to date have been in 1970, 1971, 1976, 1977, 1979, 1981, 1982, 1983, 1984, 1985, 1986 and 1987. Porsche also has an unrivalled record of over 116 outright wins in the World Sports Car Championship.

Porsche's previous Formula 1 and Formula 2 experience was limited to the brief period in early 1961 and 1962, since which time it had concentrated on sports car racing. The company had long taken the view that single-seater racing, and in particular Formula 1,was too far removed from its product and therefore the technological advantage of participation too small. Certainly, all of the road racing and sports cars described in the previous pages were powered by production based engines, with the exception of the 904 and the 917, and even in those examples engine experience has benefited the road car. The company's attitude to Formula 1 is dictated by reason, logic and a keen sense of the costs and the benefits of development opportunities. However, the chance to build a Formula 1 engine as an outside contract was not to be missed.

Thus started the beginning of Porsche's busiest and most successful period in motor sport. Besides Group C, Porsche signed a contract with TAG Turbo Engines on 12th October, 1981, to build a Formula

1 engine. The FISA Formula required a 1.5-litre turbocharged engine and McLaren designer, John Barnard, had already produced a detailed specification covering external dimensions, power and torque characteristics for the engine to be used in the McLaren chassis. Within a year a V6 engine was running on a Weissach test bed, and track testing began early in 1983. Given the cost of developing Formula 1 components, it was decided to concentrate all development upon race engines and not to spread the cost to cover qualifying engines. The rest is history.

The TAG Turbo TTE-PO1 engine built by Porsche first raced towards the end of the 1983 season, with modest results, although by the season's close it began to display the expected Porsche reliability. In 1984, the McLaren MP4/2 chassis powered by the TAG Turbo engine won twelve of the sixteen Grand Prix races and gave Niki Lauda, appropriately an Austrian, his third World Drivers' Championship and McLaren the Constructors' Championship with a record points total. In 1985, Alain Prost, Niki Lauda's teammate, won the World Drivers' Championship for the first time and McLaren again won the Constructors' Championship. In the 1986 season, Prost became the champion again. 1987 was the swansong for the TAG Turbo built by Porsche. In total, out of 64 Grand Prix entered, the TAG Turbo won 25, representing a 39.06% victory record.

The fuel consumption trials set by FISA from 1984 onwards, undoubtedly generated a great deal of effort towards efficient fuel consumption. The TAG Turbo engine had a revised Motronic manage-

1–2–3 at Le Mans. Porsche 956 cars took the first three places in the 1982 Le Mans 24-hour race.

The TAG Turbo engine made by Porsche powered the McLaren Grand Prix car until 1987. Here, mechanics work on it in the Silverstone pits.

ment system in 1985, another in 1986 and in 1987 further EE-Program 'chips'. All good development material for the future and another area in which collaboration between Porsche and Bosch provided excellent results quickly applied to Porsche production models.

In 1986, Porsche instituted the 944 Turbo Cup in Germany and the series is now an international event. It is a further opportunity for Porsche and Porsche owners to benefit from race experience, another example of the growth of the Porsche family, and a further extension of the advantages of racing road cars for road development. The 250 bhp 944 Turbo is directly descended from the Turbo Cup race car.

The marque's exceptional performance in more than forty years of factory participation has been due, in very large measure, to the reliability and consistency of its cars. This reliability is hard won, and since 1966 all Porsche developed racing cars have had to undergo extensive tests on rigs and track. Among these many tests, each car has to survive a 1000 km endurance run on Weissach's own destruction track which consists of pot-holes, humpbacks and other devices to search out any weaknesses. Meticulous preparation and build quality have, as with its road cars, been a major element in its success.

The entry of other manufacturers into Group C racing has certainly stimulated the racing scene for Sports-Prototype racing cars, but much of Porsche's racing experience has, since 1948, been with either factory or private entrant racing versions of its road cars. From the earliest days of the Glöckler Porsche to the American Holbert and Penske teams and the considerable successes of the Joest, Kremer and Brun teams, private team entrants have extended the depths of Porsche involvement in the highest levels of sports car racing adding to both technical spin-off and development knowledge. Perhaps as

Superb high performance road cars that can be raced. Porsche 944 Turbos in the Turbo Cup.

important has been the Porsche family of individual owners who race, often aided if not supported, with technical back-up from the factory or the customer service available at the circuits.

It is practically impossible to relate all of Porsche's worldwide racing successes, let alone identify every technical advance which has been transferred to road car development. However, although we have detailed some of the main strands of the relationships between racing and road use, there is perhaps an overall element of vital importance; the challenge. New challenges keep the company young and competitive. Not only are Porsche engineers and technicians gaining from racing, as with outside contracts at Weissach, it adds to their learning.

There are two further reasons why Porsche goes racing. Firstly, it instils a philosophy of getting a job done on time. In motor racing there are no excuses. If a car is to start it must be ready on time. Secondly, racing underlines the requirement for zero defects, 'to win you first of all have to finish', is a particularly apt saying. Since so much Porsche racing is in the hands of private entrants, the race and production-based race cars that are the main basis of current Porsche activity provide a vast array of development information. It must be hoped that Porsche long continues its mission, of building superb high performance road cars, that can be raced.

CHAPTER 13

THE WAY AHEAD: FUTURE PORSCHES

Since 1948, Porsche has established itself as the most innovative producer of high performance road and racing cars in the world. The company's record in motor racing is unsurpassed, from one make amateur races to twelve outright wins at Le Mans, and many other major events in the world's motor racing calendar.

The company has produced only three completely new models since 1976 plus the most advanced vehicle ever produced by any manufacturer, the Porsche 959. Only 250 of these vehicles will be produced, including those made for development and competition purposes. This automotive vehicle, the word 'car' seems rather too simple to apply to the most advanced production car in the world, was initially the victim of FISA rule changes, more of which later.

Technically the whole is, according to those who have driven it, greater than the sum of its parts; but the whole is more than worthy of the extravagant adjectives that have been applied to it. During our visit to Weissach, such was the reputation of this machine that the effect of seeing several test vehicles parked around was one of surprise; surely they should be in a special place. But, in true Porsche tradition, the future was there, never very far from influencing current practice.

The latest ultimate car from Porsche can be traced back to the four-wheel drive 911 Studie of the 1981 IAA Frankfurt Motor Show, and the Gruppe B study unveiled at the same venue in 1983. The latter was conceived to provide Porsche's racing customers with a less expensive alternative to the Group C racing car, the 956.

During the development and production stages Porsche never, of course, left the competition arena and outright success in the 1984 Paris-Dakar Raid in a 4-w-d 911 was followed up with another Paris-Dakar Raid win in 1986 – this time with a 959 – and a seventh place overall at Le Mans in the same year. The latter car was a 961 (race version of the 959) and, due to FISA's decision to drop Group B racing, it ran in and won the IMSA-GTX class. The production story took over.

The original plan had been to build and sell 200 cars required for homologation by June, 1985. Then, in 1986 the decision by FISA to ban all Group B cars removed the initial reason for the car but

Porsche 959 rally car. The winner of the 1986 Paris-Dakar Raid.

Porsche decided to proceed with design development. The Gruppe B proposal had rapidly turned into another 911 Turbo. Therefore, as happened with the 911 Turbo, it was decided that the 250 to be built for sale would be luxurious, high performance road cars which would be the (current) ultimate expression of Porsche engineering and design.

The planned production start date of 1985 was delayed for fifteen months, for a number of reasons. Firstly, the development team had, at least to some extent, been overtaken by the priorities of further engine development on the water-cooled engines. Secondly, with a production run of only 250, the suppliers of the 959 components were finding it difficult to give the project priority against other, rather larger demands. Thirdly, the intense technical complexity of the car meant that Porsche was determined to optimise the specification to provide customers with the latest in state-of-the-art technology.

The body shape bears favourable comparison with any of the most exclusive cars, of this or any other generation, and is a magnificent demonstration of the principle of Porsche engineering refinement through constant development. The unique eloquence and timelessness of both 'Butzi' Porsche's original design and this Porsche development principle is proven by the brilliance of the 911 body lines which restyling has given a futuristic shape.

The technical specification of the 959 is a roll call of all the very latest and proven (but not by many other manufacturers) technological developments in the automotive and allied fields. The external skin is fully aerodynamic, and the Kevlar based aramid and aluminium body panels are secured onto a hot dip, galvanised monocoque. The external dimensions are slightly wider than the 911, and the drag coefficient is down from Cd 0.39 to Cd 0.32 with different side, front and rear treatment. The bonded screen and the headlamps are flush with the bodywork, the former behind glass covers. The aramid engine cover has a very efficient but sympathetically designed rear wing which contributes to the aerodynamic efficiency of the total body shape; lift is zero at 200 mph. But a more important indication of the aerodynamicists' success, is that the downforce is equal on all four wheels at high and low speeds. Total weight of the car is some 3190 lbs.

Under this remarkable Kevlar coated skin, the car's absolute technological advance continues. The suspension is more akin to that of a Formula One Grand Prix car, with upper and lower wishbones front and rear, plus coil springs and electronically or manually adjustable Bilstein dampers. The wheels and tyres have a pressure loss warning system, and the tyres are 235/45 section front and 255/40 rear. Naturally, ABS is fitted and the stopping power of the 959 is literally phenomenal.

Basis of the power unit is the flat-six, four-valve per cylinder, water-cooled head and air-cooled cylinder barrels, racing engine from the 956/962 cars, with a number of important differences. The unit has twin KKK turbochargers, one of which operates from 1200 rpm to 4000 rpm, when the other comes in sequentially and they operate at a boost pressure of 2 bar (28 lbs/sq inch). The engine revs. up to 7600rpm and the maximum boost at peak torque is 369lbs/ft; indeed the engine produces over 300lbs/ft of torque from 2200rpm to the rev. limit. The 0-60mph acceleration time is under 4 seconds and top speed is quoted at 197mph.

The drive and transmission arrangements are as advanced as the rest of the car, with an infinitely split front to rear power variable four-wheel drive system. The system can be controlled electronically or with partial manual override. The manual control for the driver is a simple four position lever for dry, wet, snow/ice and off-road conditions. The Weissach engineer's advice is that the electronic control will beat any driver for effective control. The lever therefore appears to be a means for the driver to compare the excellence of the electronics! The limits of the power transfer are from 20:80 front to rear to 50:50, a final sophistication is the ability to transfer all of the power to the front wheels, if there is a total loss of rear-wheel traction.

The driver has the opportunity to alter ride heights with three settings, controlling height regardless of load. If the driver selects anything but the lowest, standard height the system will automatically reduce height to the lowest above 100 mph, and will not re-select the higher setting until speed drops below 93 mph. Such compromises as anti-roll bars are unnecessary, as the shock-absorbers are

Porsche 959, the ultimate Porsche.

hydraulic and inter-connected, electronically of course.

As with the steering on the front-engined, Transaxle cars, the power assisted system retains 'feel' and always gives the driver contact, but on the 959 they have used advanced materials including hard-chrome polished steel and Teflon sealing rings to eliminate friction, and maximise 'feel'. In all cars, from the mundane to the magnificent, the proportion of electronics in the total structure has been increasing, but in the Porsche 959 it is high, even by Porsche standards. It is all effective and contributes towards the car's unrivalled

The Porsche 960.

performance, rather than simply providing pretty fascia displays.

A further development which shows design possibilities, rather than intentions, is the Porsche 960, which was shown at the 1981 IAA Frankfurt Motor Show. This car, based on the 995 project, featured super lightweight construction using advanced materials and a prototype of the PDK transmission described earlier. The whole car was nearly eight hundred pounds lighter than the standard 928 (2405 lbs versus 3197 lbs), primarily because of the HSLA steel of the structure, and design features such as wound fibreglass bumpers. The car's power unit had of course, the TOP cylinder head engine, a compression ratio of 12.5 to 1 and was very fuel efficient.

So where will Porsche's development of the automobile take drivers in the future? Technical developments appear to be certain to continue in the direction of fuel efficiency and emissions control, whilst making no sacrifices to power loss. This will be achieved primarily through lighter body and chassis weight, and more efficient engine design and management systems. Judging from the comments made in 1987, by Professor Dr. Helmuth Bott, Porsche's Board Member for Research and Development, Porsche has absolute confidence that its concept of high performance road cars have a place in the future of the automobile. Certainly, the deep reserves of active and passive

safety that all Porsche cars give their drivers, and which we described in the chapters on driving Porsches, are invaluable assets to their customers, and all who are privileged to drive their products.

The four-wheel drive 911 is scheduled to appear in right-hand drive form in the near future. Further applications of the technical advances made in race and rally car development will appear on production models, and it is also inevitable that at least some of the technological wizardry of the 959 will as well. The 959's very sophisticated four-wheel drive, and the wheel pressure loss systems are, perhaps, two early possibilities. In the longer run, the Porsche PDK semi-automatic transmission system (see page 198) is a possible replacement for the automatic systems currently available. It is certain that if Porsche knows that any technical advance will increase performance, safety or driving pleasure, it will be applied.

Porsche cars are, undoubtedly, exclusive and will become more so. The company has survived and thrived in a post-war period which has plunged the depths and yet scaled the giddy heights of economic prosperity worldwide, and Porsches sell in a worldwide market. Indeed, Porsche history sometimes seems to defy the laws of economics by prospering and launching products in times deemed by outsiders to be distinctly disadvantageous; this with a product that

Pure driving pleasure. The 911 Carrera 4, the first descendant from the 959 is powered by a 3.6-litre 250 bhp flat 6-cylinder engine and incorporates four-wheel drive and ABS technology.

continually redefines the way in which the motorcar obeys the laws of physics.

The team at Weissach are the heart of future developments at Porsche, and of Porsches, but the customers and drivers will always be the engine that drives their endeavours. From the first Porsche 356 which Professor Dr. Ferry Porsche and his team built in 1948 to the present model range, Porsches have always been built for drivers who "Desired above all a car with superior handling that was swift, safe and intensely satisfying to drive".

That is at least one of the major reasons for the company's success. Another is that Porsche has always redefined the standards of swiftness, safety and superior handling which enthusiastic drivers can expect. Providing you, the driver, continue to want the finest car that automotive technology and engineering can build, with quality and performance of unrivalled standards, supplying them will continue to be the company's mission.

By continuing that tradition Porsche cars will continue to advance the art of automobile construction and redefine the meaning of "Driving in its purest form".

Bibliography

IAN BAMSEY *Porsche Carrera 6–962* Tenorhart Ltd 0-947914-00-5
IAN BAMSEY *Porsche 917. The Ultimate Weapon* Haynes Publishing Group 0-85429-605-0
JOHN BENTLEY/DR. F. PORSCHE *We at Porsche* Haynes Publishing Group 0-85429-228-4
LOTHAR BOSCHEN/JURGEN BARTH *The Porsche Book* PSL 0-85059-559-2
LOTHAR BOSCHEN/JURGEN BARTH *Porsche Specials* PSL 0-85059-802-8
MICHAEL COTTON *Porsche Progress* PSL 0-85059-928-8
MICHAEL COTTON *The Porsche 911 and derivatives; A Collector's Guide* Motor Racing Publications Ltd 0-900549-30-0
MICHAEL COTTON *Porsche 911 Turbo* Osprey 0-85045-400-X
RICHARD V. FRANKENBERG *Porsche – The Man and his car* G T Foulis & Co Ltd 85429-090-7
RICHARD V. FRANKENBERG with MICHAEL COTTON *Porsche: Double World Champions 1900–1977*
PAUL FRÈRE *The Porsche 911 Story* PSL 0-85059-605-X
PAUL FRÈRE *Porsche Racing Cars of the 70s* PSL 0-85059-442-1
CHRIS HARVEY *Great Marques Porsche* Octopus Books Ltd
DENIS JENKINSON *Porsche – Past and Present* Gentry Books Ltd 0-85614-083-X
DENIS JENKINSON *Porsche 356* Osprey 0-85045-363-1
KARL LUDWIGSEN *Porsche – Excellence was Expected* Automobile Quarterly Books 0-525-10117-9
MIKE MCCARTHY *Porsche* Bison Books 0-86124-249-1
DOMINIQUE PASCAL *Porsches at Le Mans* 0-85429-457-0
DR. ING H. C. F. PORSCHE AG *100 Years of Porsche – Mirrored in Contemporary History* Dr. Ing H. C. F. Porsche AG
JERRY SLONIGER *Porsche 924 928 944* Osprey 0-85045-776-9
JULIUS WEITMANN. Revised & edited by MICHAEL COTTON *Porsche Story* PSL 0-85059-673-4
JULIUS WEITMANN/RICO STEINEMANN *Project 928* Motorbuch Verlag 3-87943-518-9